SOTHIS PRESS

Sothis Press aims to completely reimagine the public domain book with an emphasis on universal access to captivatingly designed print editions. We believe that technology will enable every single book ever published to be perpetually accessible in beautiful print form. Sothis is beauty in ubiquity.

www.sothis.press
1 Waterhouse Square, London, ECIN 2ST

Oregon Geology

A Revision of "the Two Islands"

Thomas Condon

SOTHIS PRESS

ISBN 978-1-0368-0356-8

This edition is published by
Sothis Press © 2023. The company
is registered in England and Wales.
The Company Number is 14900389.

OREGON GEOLOGY

A REVISION
OF
"THE TWO ISLANDS"

BY

THOMAS CONDON, Ph. D.

Professor of Geology, University of Oregon

WITH A FEW TRIBUTES TO THE LIFE AND WORK OF THE AUTHOR

EDITED BY

ELLEN CONDON McCORNACK

Portland, Oregon

The J. K. Gill Company

1910

Printed by
The Irwin-Hodson Company
Portland, Oregon

TABLE OF CONTENTS

ILLUSTRATIONS

PREFACE TO THE SECOND EDITION

Since it has seemed best to the friends of this volume that in the second edition the original title be changed, a word of explanation seems necessary lest the many friends of the author fail to understand the innovation. There have always been those who felt that the title "Two Islands" might so easily be mistaken for that of a volume of romance or adventure as to seriously interfere with the distribution and usefulness of the work. But more important still, the original title places the greatest stress upon that link which, in the light of later exploration, has proven the most vulnerable point in the whole chain of reasoning. For it has been questioned whether the two oldest portions of land in Oregon were literally geographical islands. By referring to a publication* of the United States Geological Survey it will be seen that Doctor Diller has mapped the Siskiyou region as an island surrounded by a Cretaceous sea. But the island character of the Shoshone or Blue Mountain region has been seriously questioned.

*Note. Geology of the Lassen Peak District.

PREFACE TO THE SECOND EDITION

If the reader will turn to any map of the United States he will find the Wasatch Mountains as part of the eastern boundary of Idaho, and if he is familiar with the mammalian life of the Eocene period, he will remember that Professor Marsh and others have described a wonderful and varied fauna of large mammals which lived on the borders of an old lake east of the Wasatch Mountains. In fact the mountains themselves formed the lofty western shore line of that Eocene lake. But of all this abundant life, not one well identified fossil mammal has been found in Oregon belonging to that Eocene epoch. The author of The Two Islands reasoned that the Oregon land must have been cut off from the Wasatch land by an intervening body of water.

The apparent absence of these fossil mammals from our Oregon Eocene is still an unsolved problem. Some authorities believe that the mammals beyond the Wasatch did live in Oregon, and think their fossil remains may yet be found. Therefore it seems best that this controverted question be not unduly emphasized by retaining the original title of the book. Fortunately this narrative of Oregon's geological growth, with its rich store of scientific facts and its vivid picture of Oregon's past, is not dependent upon the very subordinate question of whether the Land of Shoshone was in Cretaceous times entirely surrounded by water, or whether it was a bold and lofty headland jutting out from the Rocky Mountains. If one region be an island and the other a peninsula, still, the original treatment of the two sections, as given by the author, would remain unchanged, for the dynamic forces that produced them, the rocks of which they are composed, and the life upon their shores, were for ages virtually identical.

ELLEN CONDON McCORNACK.

PREFACE BY THE AUTHOR

The want which these pages attempt to supply is a popular rather than a scientific one.

For years our general government has been publishing, through railroad surveys and the annual reports of the United States Geological Surveys, a large mass and wide range of geological information on the structure and history of our Western coast.

But this large body of information is so scattered that few have the time to collect enough of it to form a continuous unity of its history. Besides, there are many things in the geology of Oregon of lively interest to the young and the uninstructed, and running through them all are threads of a continuous unity that seems capable of a possible narrative form such as might increase the interest of the young.

An attempt to meet this double want, not with a fresh contribution to science, but with an attempt at picture making for the uninstructed, has led to the writing of these pages.

THOMAS CONDON.

University of Oregon, Eugene, 1902.

A SKETCH OF THE AUTHOR'S LIFE

There was a limestone quarry near the home of Mr. Condon's childhood that must have made a deep impression upon his thoughtful mind, and shed the affectionate glamour of early association over his study of the rocks, for his interest in geology began with his childhood.

Fortunately for him his family left the old home in southern Ireland, and crossing the Atlantic, made their home in the city of New York. Here we find the future scientist an active wide-awake boy, full of life and with a strong appetite for knowledge. Some of his leisure hours were utilized in exploring the old Revolutionary fortifications near the city. And occasionally he spent a half holiday hunting rabbits in the wilds of what is now Central Park. A few years ago, in speaking of those days of his boyhood, he referred to his study of algebra and then said: "But when I took up geometry, it lifted me to the clouds. I drank it in as a mental food." It seemed to be the pure, beautiful logic, the perfect chain of reasoning, that appealed to his mind.

At about eighteen years of age, he was working, studying, and teaching in Camillus, Skaneateles, and other places in central New York, where he finally entered The Theological Seminary at Auburn, while teaching in the evening school at the state prison there. The history of those years in the lake country of central New York would read like a romance of extreme interest. But in spite of all difficulties, he spent many leisure hours among the hills and quarries, gathering fossils and studying the geological formation of the region.

But he had heard of the Whitman Mission in the far west, and had made up his mind to go as a home missionary to the Oregon country, and in 1852, with his young bride, he sailed in a clipper ship around Cape Horn for San Francisco. After a long and eventful voyage, they found themselves in the newly settled and unexplored Oregon.

Trappers had long known it as a land of furs; miners had known it as a land of gold; the early pioneer had found it a country with rich and fertile soil; but its scientific resources were still undiscovered. The questions that had dawned dimly upon his mind as he played by the stone quarry of his childhood, the questions that were kindled into life as he studied the fossils of central New York, the questions of the how and wherefore of creation must have come to him with new force as he looked out upon the fertile valleys, grand mountains and noble rivers of his new home.

But the activity of these first years left but little time for scientific research; for new homes must be built, land cleared, crops planted, schools started, churches organized, and hostile Indians subdued, and there were but few of these labors of pioneer life in which he did not take an active part.

After ten years of life in western Oregon, Mr. Condon, wishing for a more needy field, moved his family to The Dalles, then the head of navigation on the Columbia, the gateway through which all the rough, reckless mining population must pass on their way to the newly discovered gold fields of eastern Oregon. Here, too, was an army post from which men and supplies were sent to all parts of the northwest.

An army officer returning from an expedition against hostile Indians, brought Mr. Condon his first eastern Oregon fossils, from the Crooked River country. These fossils aroused the keen interest of the student of nature, and in 1862 or '63, he obtained permission to accompany a party of cavalry carrying supplies to Harney Valley. They returned by way of old Camp Watson, on the John Day River, and here Mr. Condon found his first fossils, in the now famous John Day Valley.

These glimpses of this fossil field only served to make him eager for more, and as soon as the Indians had been subdued and it was safe to venture among those hills and ravines without an army escort, Mr. Condon spent his vacations exploring in the John Day country. On one of these trips he found and named Turtle Cove, which has since proved to be one of the richest fossil beds in the valley. He employed young men to spend their summers collecting the fossils exposed by the wear of winter storms. He made friends with the rough teamsters who drove the great government freight wagons from Fort Dalles to the army posts in the wilderness. As these teamsters returned with empty wagons they often brought a few rocks or a fine box of fossils for their new friend at The Dalles. In a few years Mr. Condon found in his possession a large quantity of valuable material that must be classified and described. But he was without scientific books, was thousands of miles from the great libraries and museums of the east, and far from other scientists with whom to confer.

Fortunately at this time the United States Government was making its famous geological survey of the fortieth parallel, embracing a strip of land one hundred miles in width, and connecting the geology of the great plains east of the Rocky Mountains with that of California and the Pacific Coast. One evening as this great work was nearing completion, Mr. Condon was delighted to learn that Clarence King, the leader of the survey, had reached The Dalles, and he lost no time before meeting this distinguished geologist. Mr. King was deeply interested in the pioneer discoverer's account of Oregon geology, and the next day found him in the Condon home, studying the unique collection.

A SKETCH OF THE AUTHOR'S LIFE.

Not later than the spring of 1867, Mr. Blake, an eastern geologist, visited the cabinet at The Dalles, and on his return voyage carried with him a few specimens of fossil leaves originally from Bridge Creek in the John Day Valley. These were perhaps the first Oregon specimens to find their way to the Atlantic Coast. They soon fell into the hands of Dr. Newberry, of Columbia College, New York, who, being a specialist in fossil botany, longed earnestly for more. After talking with Clarence King, in Washington, and learning from him more of the Oregon geologist and his country, Dr. Newberry wrote Mr. Condon in 1869, and received in response a box of fossils of which he writes: "I received your two letters with great pleasure. Since then the box has safely come to hand and that has given me still greater satisfaction, for I found it full of new and beautiful things which fully justified the high anticipation I had formed judging from your letters and the specimens brought by Mr. Blake."

In the autumn of 1870, Arnold Hague, also connected with the geological survey of the fortieth parallel, spent a month in Oregon, part of the time being at The Dalles in discussion over the geological problems of the Columbia River region. That this visit was a source of mutual pleasure, is shown by a subsequent letter in which Mr. Hague refers to his "month in Oregon in 1870 as one of the pleasant memories of the past."

But a new era was dawning for "the Oregon country." The first transcontinental railroad had touched the Pacific and with it came many large parties of cultured tourists who, wishing to look upon the grand scenery of the Columbia, found themselves obliged to spend the night in The Dalles. In this way it often happened that late in the afternoon, a party of fifteen or twenty ladies and gentlemen would gather at the home of the Oregon geologist and spend a pleasant hour studying the life of past ages.

In 1870, Mr. Condon shipped his first boxes of specimens to the Smithsonian Institute at Washington, and from there they were sent to Dr. Leidy, of Philadelphia Academy of Sciences, for expert examination. The National Museum was glad to receive these new fossils from the Pacific Coast and promised its official assistance in every way possible.

A few months later, of this same year, Professor Marsh, of Yale College, wrote from San Francisco as follows: "I have heard for several years a great deal of the good work you are doing in geology and of the interesting collection of vertebrate fossils you have made, and I intended during my present visit to the Pacific Coast to come to Oregon and make your acquaintance personally and examine your fossil treasures which my friends, Professor George Davidson, Clarence King, Mr. Raymond, and others have often wished me to see." And a little later, Professor Marsh writes urging that all fossils of extinct mammals be sent to Yale to be used by him in a work on paleontology, gotten out by the United States Government in connection with the survey of the fortieth parallel.

During these years, many Oregon fossils found their way to the educational centers of the east. If they were fossil leaves, they were sent to Dr. Newberry, of Columbia College; if shells, to Dr. Dall, of the American Museum of Natural History; if fossil mammals, to the Smithsonian, or to Marsh of Yale or Cope of Philadelphia. A few of these were sold, some of them were sent in exchange for eastern fossils, but most of them were simply lent in order that they might be classified and described by scientific experts.

In May, 1871, Mr. Condon published in the Overland Monthly, his paper on "The Rocks of the John Day Valley." And in November of that year his article entitled, "The Willamette Sound" appeared in the same magazine. The latter was perhaps his favorite of all his geological writings. He felt that "The Rocks of the John Day Valley" might need revising after a more thorough exploration, but that "The Willamette Sound" would endure. Both of these papers are given in "The Two Islands," published in 1902.

Dr. Diller, of the United States Geological Survey, has virtually accepted "The Willamette

Sound," and incorporated its substance in his report of the geology of northwestern Oregon, his only criticism being the suggestion that the waters of the sound were probably even higher than noted in the original publication. These two papers fairly represent Mr. Condon's strength as a constructive geological worker. They indicate his ability to begin at ocean level and by means of mountain upheavals, marine and lake sediments, fossil leaves and bones, and volcanic outflows, to reconstruct and make wonderfully vivid the geological past of a new country.

From this time on, the sense of lonely isolation that had so hampered him in his work, gave place to the most cordial intercourse between the Oregon pioneer and distinguished scientists of the United States and Canada. In 1871, Mr. Condon had the pleasure of showing Professor Marsh and a large party from Yale College through the new fossil field, and a little later Professor Le Conte of the University of California was introduced into the same John Day Valley. The latest scientific publications began to find their way into Mr. Condon's library in exchange for information and material freely given to eastern workers. The stimulus of all this stirring intercourse by exchange, correspondence, or personal conversation with some of the most learned men of the age, was a great boon. Life in the strength of his manhood was full of buoyancy and joy, a grand opportunity for usefulness.

It gave Mr. Condon real pleasure to sit down beside a rough block of sandstone with only the corner of one glistening tooth in sight, to pick and chip and chisel until another tooth and part of the jaw were seen, to continue with careful skill until the beautiful agatized molars were laid bare, to work patiently on until there stood before him, no longer the shapeless mass of stone, but a fine fossil head to add its testimony to the record of the past. But it gave him greater pleasure still, to work with rough, unpolished human character and discover the glint of gold hidden under the rough exterior. The book of nature was indeed fascinating, but did not appeal to him as did the work with men. He had the artist's eye for seeing the beautiful in character and the enthusiasm of a sculptor for shaping rough, faulty human nature until its beauty reflected the Divine.

To many minds, these two lines of interest, the development of character and the study of nature, would seem incongruous, but to him they were both God's truth, the one the preparation, the other the culmination of God's work. And yet, strange and unusual as is this combination of geologist and minister, it seemed exactly what was needed to equip one for usefulness thirty or forty years ago. For these were years of great stir in the scientific world.

A SKETCH OF THE AUTHOR'S LIFE.

The author of "The Origin of Species" and "The Descent of Man" had given his theory of evolution to the world. The grand truths developed by that galaxy of brilliant English writers, Spencer, Huxley, Tyndall, and others had been seized by materialists who were calling upon all thinkers to discard the Bible as out of date because not in harmony with scientific thought. Christian ministers were not scientists, and the principles of "Higher Criticism," if thought of at all, were considered dangerous heresies against which to warn their people. To Mr. Condon, the theory of evolution presented to the human mind a wider conception of God than the world had ever known. It involved a plan of unthinkable grandeur; beginning with the smallest, simplest things, gradually unfolding into more complex life, often interrupted by some great upturning of nature, but never losing the continuity of purpose, the steady progress toward the culminating glory of all: the spiritual life of man.

To have all this new wealth of spiritual vision appropriated by materialists was a source of deepest sorrow. The storm, starting on the intellectual heights of Europe, was slowly traveling westward. A little later, magazines were full of the subject and materialism was creeping into college life with the claim that evolution was antagonistic to religion. The young men who studied science found few leaders so endowed as to interpret the beautiful adaptation of the doctrine of evolution to the spiritual life.

Mr. Condon saw that the old ramparts erected by theologians were no longer a safe retreat; that the church must be defended even by science itself, and he longed to help unfurl the Christian banner over this newly discovered realm of truth. He felt his most effective work could be done with his cabinet in shaping the immature minds of Oregon's sons and daughters. This, with the growing educational needs of his family, finally led him in 1873 to take his place with the faculty of Pacific University at Forest Grove, and later in 1876 to accept the chair of Geology and Natural History in the State University.

In 1876, shortly after reaching Eugene, Mr. Condon, in company with a son of ex-governor Whiteaker, made a trip to the Silver Lake country in southeastern Oregon. Here they gathered a fine collection of beautifully preserved fossil bird bones, which were sent east to be described, but seemed too rare and valuable to be returned, for, in spite of many efforts to recover them, they were finally lost to the rightful owner. Fortunately, they had been previously examined and described by Dr. Shufeldt, an expert in the study of fossil birds, and to him we are indebted for much interesting knowledge of the ancient life of the region. This same locality has also yielded some of the finest specimens of fossil mammals in the state.

By this time, Oregon had passed out of its pioneer stage and was looking to a broader expansion of statehood, with all its hidden possibilities of industrial development. Men were asking, "Have we coal in Oregon?" "How shall we utilize our gold-bearing black sands?" "Have we the right geological formation for artesian water?" "Have we cement rock, copper, or limestone?" Letters on all of these and many other problems kept coming to Mr. Condon from near and from far. These questions and the investigation necessary for their answers resulted in his acquiring an extensive knowledge of the industrial problems of the state. If any one wished to bore for artesian water his advice was asked. The discoverer of a fresh prospect for coal, copper, asbestos or marble, must send him a sample specimen and ask his opinion of its value, and he was always ready with a word of advice, a bit of encouragement or a needed caution.

All these years he had been glad to share his rapidly increasing knowledge with the people of the northwest. The old river steamers and slow moving trains of early Oregon often carried him to fill lecture engagements, and he was usually cumbered with many heavy packages of specimens and choice fossils to illustrate his subject. Sometimes the lecture would be before a cultured Portland audience; sometimes it was a course of lectures for some growing young college; or perhaps a talk to the farmers at the State Fair, upon the formation and composition of soil. But as the years passed, most of his time and strength were given to his teaching at the University, while his summer vacations were spent with his family at his Nye Brook Cottage by the Sea.

Here his life was almost unique, but it again brought him into the most friendly relations with many classes of people from all parts of the northwest. Sometimes there were formal lectures before a summer school, but more often there was an informal announcement that "Professor Condon would lecture on the beach," perhaps near Jump-off Joe. And here his audience would gather around him in the shelter of the bluff or headland, some standing, some sitting on the rocks, others perched upon the piles of weather-bleached driftwood, while the children sat "Turk fashion" upon the dry, glistening sand. And he, with his tall alpenstock in his hand, his broad hat and loose raglan coat, made a picturesque figure standing in their midst. Perhaps he talked of the three beaches, the one upon which they stood, and the two old geological beaches so plainly visible in the ocean bluff behind them. The banker, the college president, the physician from a distant part of the state, the young city clerk, the carpenter, the teacher of the country school, the farmer and his family taking an outing by the sea, even the high school boy, and the children, all listened with interest. And when the talk was over and all their questions had been answered, the motley gathering strolled leisurely away. But the rolling breakers at their feet, the hurrying scud and blue summer sky, all had a new significance as they pondered on the mystery of creation.

Or perhaps a geological picnic was planned up the beach to Otter Rocks. After a brisk ride of a few miles over the hills and along the beach, Mr. Condon's carriage would stop, the other vehicles would group themselves around near by, and, standing in his conveyance, he would give a short talk on the geological formation of the particular cove or headland with its base of old sandstone full of fossil shells. Then the company would move on, and after a few more miles of delightful beach ride upon the hard sand near the breakers, they would leave their carriages, gather their picks, hammers and chisels, and spend an hour chipping fossils from the bluff or from the large bowlders at its base. The next stop would be to lunch near Otter Rocks and explore the unique Devil's Caldron or Punchbowl and the interesting beach beyond.

But the most common picture, the one that must make the Condon Cottage at Nye Brook an almost sacred spot for some, was the party strolling homeward from a morning on the beach—especially at low tide. They always stopped beside his cottage door to show their treasures to Mr. Condon. There were baskets, tin pails, and all sorts of packages filled with curios gathered on the morning walk; one had a rare shell-fish, another an unusual fossil, some had sea moss, others only a group of bright pebbles, while a few proudly exhibited their water agates. All had their eager questions and his kindly, helpful interest never failed; for if some child but left his cottage door with eyes large and shining with a new joy, because it had caught a glimpse of the beauty of knowledge, he was content. And so his summers passed.

Meanwhile he had been carrying on his original research work by taking trips to the southwestern part of the state, and was slowly filling out his geological map of Oregon.

Mr. Condon's love for knowledge was not confined to natural science, for his interests were broad as the universe. To him, human history began with the men of preglacial age, and he sought eagerly for every ray of light that archeological research could throw upon the old Cave Dwellers of prehistoric times. He studied all primitive peoples, their religion, industries, and social development, and endeavored to trace their relationship to common ancestry. There were but few obscure nations of the world in which he was not deeply interested; he knew their past history, their present political condition and struggles for liberty. He prized the history of our Aryan ancestors and treasured their old Vedic hymns as among the first bright glimpses of the human soul in reaching out for its Creator. The religion, art, and literature of the Egyptians, Arabians, Persians, and Greeks were to him a source of great pleasure. He followed the lives of noted statesmen and was most enthusiastic in his admiration for the world's true heroes. All great religious movements, including the higher criticism and the relation of science to religion were matters of absorbing interest. And yet there were but few who knew and loved Oregon's trees, shrubs, and wild flowers so well as he.

In 1902, after passing his eightieth birthday, Mr. Condon published his "The Two Islands," a popular work on the geology of Oregon, which, aside from its scientific value, will be prized for its clearness and simplicity of style and the subtle charm of his own personality as constantly revealed in its pages. It was not written for technical scientists, but for the larger circle of readers who love to catch such glimpses of the progress of creation. No, Mr. Condon was not a specialist, either by nature, inclination, or education. And it was well for the early development of Oregon that he was a true pioneer with a large appetite for all knowledge, a keen pleasure in imparting that knowledge to others, and a broad, sympathetic outlook into the needs of the Northwest. If he had been a specialist, be might have received more technical credit in the scientific world, for he discovered many new fossils and named but few. But what is the naming of a few fossils more or less, when compared with the grandeur of such a broad sweep of knowledge, permeated by such a beautiful spirit of helpfulness?

The pioneer work in this new and unexplored state, so remote from the great centers of learning, required just his type of mind; just his habit of first sketching in the broad outlines and then filling in the details with all their picturesque beauty. For as the artist works, he worked. A colleague who wrought by his side has said of him, that instead of beginning with the minute details, and progressing towards the large facts of life, he always began with the broad outlines, the great principles of any subject, and worked down to its details.

After this active, eager life had passed and failing health gave him ample time for retrospective meditation, he realized that he had lived through a grand period of pioneer history and remarked, as he looked forward into the future in store for the rising generation, "I do not know that I would exchange the rich chapters of my own life for all the future opportunities of these young men."

For he was the pioneer geologist who, by his own original research, caught the first glimpse of Oregon's oldest land as it rose from the ocean bed; he saw the sea-shells upon her oldest beaches; watched the development of her grand forests: saw her first strange mammals feeding upon her old lake shores; he listened in imagination to the cannonading of her first volcanoes and traced the showers of ashes and great floods of lava. He followed the creation of Oregon step by step all through her long geological history and then entered with enthusiasm into the industrial and educational development of her present life.

But above all, infinitely above all, he prized and labored for the noble character of her sons and daughters. Is it any wonder that his heart was full of gratitude to God for having guided him into such a rich heritage of life?

ELLEN CONDON MCCORNACK.

CHAPTER I.

THE STONE QUARRY.

The ancient traveler and historian, Herodotus, long ago put on record the statement that he had heard from Egyptian priests that in a range of hills along the eastern border of their country the rocks contained buried sea shells. The priests contended that these shells were in such positions and numbers as to indicate that in some countries the sea and land had changed places.

The same thought occurs today to many an humble quarryman who disclaims all thought of geology, but who does know his stone quarry is nothing else than a petrified sea beach. Professor Leslie finely says, "Every rock fragment that lies upon the surface of the crust of the earth has legibly written on it and around it, the facts of its history if we would only study them." The evidence for our study of these facts varies all the way from the vague conviction of the quarryman's judgment to the convictions that are founded on the closest scrutiny of the facts by men whose whole lives are devoted to the study of just such things with all the skilled culture of the age to help them. In addition then to the quarryman's conviction, we have skilled students of science testifying that they find in these very stone quarries the materials for the exact study of plant and animal life. So complete are these materials that they satisfy all inquiry and produce a conviction that in prying apart the stone layers of the rocks, the scientist is in reality opening the leaves of the past history of our world; and that in these buried leaves of sandstone and mud deposits of seas and lakes of former ages, he is uncovering not only the real shells and bones of the life of the period, but even the ripple marks of the waters that once covered them.

On Plate I. we have a very good engraving from a photograph of one of these rock fragments of an ancient sea beach. This fragnment here photographed is twenty-two inches in length and sixteen inches in breadth. The twelve or fifteen large shells of the print are of the family Mactridæ, one of which is so well represented in this tablet of stone is the genus Mulina, a form that was very abundant along the coast when this beach was formed. A handsome elliptical shell, the Tellina, is represented in the plate by eight or ten impressions, some of which are distinct enough to give a good idea not only of the shell form, but revealing the minute lines of its growth or its ornaments. Two or three straight impressions are made by the fragments of the genus Solen, a razor clam which was an abundant shell on the Pacific Coast of those times. In addition to these bivalves, the

Fragment of Miocene Sea Beach. **Plate I**

١

reader may notice three or four snail like shells of the family Naticidæ. These two were abundant on those ancient sea beaches of which this piece of rock is a fragment, having been taken from a ledge of sandstone in the foothills of the southern part of the Willamette Valley. It will be noticed that the surface of this specimen is thickly strewn with those shell forms varying in size from a handbreadth to the minutest specks. These lie exposed on the surface or buried in the mass of the rock at different depth; some so slightly held that the shell material is easily displaced, others almost buried from sight, still others only found by making new fractures. Many of the bivalves are tightly closed as when alive, others spread wide apart and all these covered with sand and mud just as one sees such forms along a sea beach of today.

And now of all these it may be truly said that they are not strangers to the student of today, but forms familiar to all who have even a slight acquaintance with the ocean life of our own times. As soon as one accepts the conviction that this piece of rock is a fragment of an ancient sea beach then at once follows the question, "how far does this old fossil beach extend?" One is compelled to answer, "thousands of miles like our present beaches." And this leads to another inquiry: "Are these other beaches older or newer than the one represented by this tablet?" Many are older and others newer, the newer ones where they touch it always overlying, the older ones underlying the beach of this tablet; and no part of the continent but possesses its share of this wonderful history of the past. The Rocky Mountains in their highest reaches bear upon their shoulders these records of a former life under the seas.

Let us look at the subject from another point of view: If we make a rough catalogue of the stone-quarries of western Oregon, including in our list only those that have been worked sufficiently to give a fairly full set of their fossils, we shall find one at Ashland, one west of Medford, one at Jacksonville, and one or two in Josephine County. In Douglas County we find a notable exposure of fossils on the North Umpqua, twenty miles east of Roseburg; in Coos County we find one at Cape Arago, another on Coos River, a third on the Coquille River. In the Willamette Valley we find twenty or more well worked stone-quarries along both lines of foothills from Eugene to Portland. Now, if we collect from all these localities their characteristic fossils, and spread them out before us and then call upon our quarryman to examine them without any help from science, he will at once separate them into three groups. Jackson and Josephine County fossils, as well as the rocks that contain them, having a family likeness, he will place by themselves in one group. He will as promptly select Douglas and Coos County rock and fossils, and make of these a second group. He will with equal readiness place all those from the Willamette Valley in another, a third group. If you ask him for his reasons he will answer, "O, they look different," and so they do.

If now we doubt the soundness of our quarryman's opinion and call in a student of science, he will at once verify the classification of the quarryman and declare that nature herself pronounces its correctness. Plainly this will at once suggest to our quarryman the explanation that these different groups of rock may be the petrified beaches of three different beach periods, and in this explanation the quarryman and the scientist are one. This new thought becomes important. Let us see what it implies. If we designate the first, the Rogue River group, because of its association with the Rogue River Valley; the second, the Umpqua group, because of its association with the Umpqua Valley; and the third, the Willamette because of its association with the Willamette region, and these three groups are held to represent the sea beaches in their relation to each other in the order of their occurrence, this order must be universal. This, also, is found to be true, that whenever these groups are found in contact, the lower one is always the Rogue River group, the Umpqua next, and the Willamette the uppermost. Not only is this true on the Pacific Coast.

THE STONE QUARRY

The world over, the group of rocks we have associated with the Rogue River Valley, geologists have called the Cretaceous; the group we have associated with the Umpqua region they have called the Eocene; the group we have associated with the Willamette region the geologists of the world have named the Miocene.

Trigonias.

Plate II

٢

CHAPTER II.

SOURCES OF MATERIALS

In the preceding chapter the reader's attention was centered on a few type specimens of the shell forms found in stone quarries. It is now proposed to dwell on the materials of the enclosing rocks in explanation of their structure and history. In order to do this more fully, the thought of the stone quarry must enlarge itself into the wider thought of the whole rocky layer of which this stone quarry is but a small fragment. In short, we are to think of a sea or lake beach, hundreds of miles in length, strewn with shells and bones, often with leaves, fruits and branches of trees, all buried in sand or mud and elevated to the crests and slopes of the hills and changed to solid rock. It is with this wider group of facts we are now concerned, and we propose to inquire about the origin of the rock materials, their travels, and the agencies that elevated them to their present positions, as well as to the hardening process that left them solid stone.

If we travel for even a few miles along the banks of any large river just after its flood season, scarcely any fact awakens more interest than the enormous quantities of mud, sand and gravel one sees left by the flood in the quiet reaches of the stream during this period of high water. In a river bed of the magnitude of the Columbia, the quantity of this annual deposit will awaken surprise when seen at any one point of its course, but to realize its full magnitude, one must think of the whole length of the river bed bordered by these deposits; and further, that they are not stationary in any portion of the river bed, but all drifting further down stream every flood time till they reach the ocean or the lake, while fresh materials are drifting from the mountains to take their place. If now we would trace this stream of sediment still farther toward its fountain head, we should find its continuance in the currents of a thousand rills made turbid by rains that wash downward what last winter's frosts loosened from rocks and ledges; and that these rills converge into brooks, the brooks into creeks, the creeks into rivers, and these into the Columbia.

Such in brief is the course of this great flood of sediment that runs a continuous stream from the mountains to the sea. Once out at sea, the current that brought it there ceases; the heavier particles drop to the bottom at the mouth of the river, while the lighter and finer materials are drifted by the currents of the ocean, till the whole has found its resting place as water sediment, covering in its muddy or sandy beds whatever of shell or bone or branch the life of the time casts off to be received and covered in these wide spread deposits.

Admitting now that all these river currents are forever washing sediments into the seas and lakes and burying in their masses the harder portions of animal bodies, such as bones, shells and teeth, there remains the fact that these are found in sea or lake beds, but not on the land. Our stone quarries are on the land, and this brings us to our next inquiry.

Are sea beds ever elevated into dry lands? Nineteen hundred years ago, Julius Caesar, at the head of the Roman army, invaded the country of the ancient Britons. Supplies for his army were brought from Italy in ships, and Caesar had a stone wharf or pier built in a sheltered bay of the Scottish coast for convenience in discharging cargo. The ruins of this pier are still visible, but they are so high above the water that it would require a change of level of over sixty feet to enable a Roman galley to reach the spot today. Here is plain evidence that since the days of Caesar, the coast of Scotland has risen from the sea over sixty feet in nineteen hundred years.

A careful series of measurements conducted under the direction of the Swedish government has established the fact that nearly the whole of Norway and Sweden is rising slowly above the ocean. This change of level has been going on for thousands of years, and has lined both the Baltic and Atlantic borders with elevated beach terraces, showing stage after stage of successive elevations, all geologically recent. The rate of this elevation is stated by government authority as amounting to five or six feet in a century at the North Cape, while less than this southward. A still more impressive group of such facts may be gathered from our own Pacific Coast. Hundreds of miles of these elevated beaches have been observed and carefully surveyed along the coast of South America; but we leave these to fix our attention upon our own home share of such world facts. For this purpose we select four points of observation along our own coast as exhibiting such recent elevated sea beds to a striking extent.

At Cape Blanco, near the lighthouse, one may see an old sea beach elevated two hundred and ten feet above tide water, in which shells, for the most part like those now living in the neighboring bays, are covered in the sand and mud deposits in which they once lived. Among the most frequently occurring of these were Schizothærus—our large blue clam—Saxidomus Nuttalli and Mya truncata. This Mya does not inhabit our waters now, but is found in Alaska, scarcely distinguishable from the fossil one of the elevated beach. Here would seem to be conclusive evidence that during the geological horizon to which these shells belong, the Pacific Ocean buried them along its beach line two hundred and ten feet above its present reach.

A similar raised beach may be seen near the mouth of the Coquille River on the hill above the town of Bandon. Here the number and variety of shells found was still greater, but at a less elevation. Schizothærus, Saxidomus and Mya, like those of Cape Blanco, were found in an elevated beach one hundred feet above present waters, and so entirely undisturbed as to leave no doubt as to their having been deposited there by the ocean.

A third one of these elevated beaches may be seen on the Yaquina coast, with similar recent shells and sea bed surroundings at an elevation of eighty feet, finely supplied with a rich variety of recent species of shells and mixed with spruce cones, evidently buried with these shells and scarcely distinguishable from the cones of living species along the coast. This elevated beach at Yaquina merits careful study. At the town of Newport, on Yaquina Bay, a fine example of these elevated beaches may be seen with all the details of its history plainly legible. The bluffs upon which part of the town is built, rise here to the height of over a hundred feet above present tidewater, seventy or eighty feet of which is marine sediment, plainly due to a broader and higher shore line of the bay of other days. In the upper part of this sedimentary portion of these beds of sand and mud, may be found imbedded marine shells many of which are scarcely distinguishable except by the stain of age, from the shells of the surrounding ocean. On the Yaquina coast one may stand in the presence of three beaches belonging to three widely separated periods. The present beach under foot is in process of building from day to day, the surf spreading the sand along the shore strewn with fragments of shells and bones that belong to the life of today, and covering these with layers of sand and mud. This working process is now occurring before the eye and making up a fresh package of geological truths from the record of today. The dark colored layer at the base of the bluff that forms the background of the view is a good illustration of the former sea beach, in this instance one of striking fitness to impress this truth because of its extent, for it is exposed along the shore for twenty miles, and in the tilted angle of its position, as well as the character of its contents, bearing out its own individuality and impressing the beholder with the conviction that it, too, is a sea beach, but of other days than these.

The more elevated portions of the bluff, as far as its stratified condition reaches, is still another sea beach, deposited by waters that stood at a higher level than those of the present time. Now, here in place are the very facts our quarryman guessed at and our scientist studied to enable him to solve the problem of the stone quarry.

Yet another of these elevated beaches, one may easily trace along the bluffs of Shoalwater Bay, on the Washington coast.

Sea beds then are often elevated into dry land, and the many instances of this change and their continued relation to each other and to the whole ocean border, prove that our stone quarries are but fragments of once continuous sea beaches.

One other inquiry remains: granted that these fossil bearing stone quarries were once the beds of lakes and seas, and that they were slowly elevated from their former level to their present positon, how were these great beds of sand and mud changed to rock? The pressure of the great weight of overlying sediments will explain part of this result; the cementing power of lime, of iron and of pasty quartz will easily explain much more, especially as all these are found contained in these sediments in the varying conditions of such cementing agents.

CHAPTER III.

OREGON'S FIRST LAND

The geological history of the Pacific Coast consists chiefly in the description of the slow elevation of successive belts of the bed of the ocean into dry land; and the progressive additions of these to the western border of North America.

These belts of ocean bed have not only been elevated into dry land, but in varying degrees hardened into rock, though retaining through all these changes the clearest evidence of their former sea bed condition. The floor of the sea from which these beds were lifted, was in favored places strewn then, as such places are strewn today, with shells, fragments of corals, or of bones,—all bearing record of the life of the period in which they were covered by the ocean sediment.

A minute and careful study of such rocky forms, by some of the best minds of the present century, has secured such results that students of geology may now speak with confidence of many great changes in the former life of the world. So carefully have the results of these geological studies been formulated, that it is entirely practicable to tell what portions of any country were first above the seas, and often to trace the successive additions

Fragment of Cretaceous Sea Beach. *Plate I*

to the land until its outline is recognized as the present continent.

Following this method in deciphering the geological history of Oregon, one is carried back to a time when this region, which we now call our home, was covered by the ocean.

In the natural world, the evident results of violent upthrusts of portions of the earth crust, are now accepted as among the trustworthy records of many lands. These disturbances were sometimes accompanied by great heat, often by violent earthquakes and the outflow of melted rock, and sometimes only by heat enough to change the materials without melting them. Oregon's geological history had its origin in such a violent crumpling of its ancient sea bed. And, when the disturbance that caused this ceased, and quiet was restored to the region, there was left as a result, two distant bodies of land. One of these occupying the eastern portion of what is now the Blue Mountain country and probably reaching into Idaho, we will call the land of Shoshone. The other extending over what is now the southwest corner of the state of Oregon, together with a portion of northern California, we will designate as the land of Siskiyou.

The violence that caused the elevation of these lands above the sea level was such that the original sediments of which they were composed were changed. If originally beds of limestone, these in the process of upheaval became marble; if originally clayey or argillaceous, the change was into hardened slate; and pasty masses of granite were forced into the more yielding portions, still farther altering the now metamorphosed mass.

It will be readily seen that whatever remains of shells, bones or leaf impressions, these older sediments may have contained in their original condition, would be marred in form, if not changed in substance, by the heat and violence through which they passed, until the record of life they once bore was dimmed in portions less exposed to violence, and in others entirely lost.

It would follow, too, that after the violence of this upheaval had passed away and quiet was again restored, marine life would slowly regain its place on these freshly formed beaches.

In tracing the early stages through which these regions passed, mention was made of changed slates and limestones and their eruptive accompaniments of the granitic rocks. So close are the points of resemblance between serpentines from Canyonville and Cow Creek in Douglas County, and those from Canyon in eastern Oregon, that hand specimens from these localities are scarcely distinguishable the one from the other. It is, too, of like import that the altered limestones of Bridge Creek, in the John Day Valley, and those from Woodville in the Rogue River Valley are easily mistaken for each other. A like resemblance is at once seen between the rich brown Gabbros from the lower canyon of Rogue River and those of the Blue Mountains east of Canyon City. Now the places these rocks occupy in the framework of these regions and the great similarity in the rocks themselves, would indicate that the same elevating forces operating at the same time and upon similar materials, gave origin to these lands.

If now this ocean outlook of ours has been sketched with anything like fidelity to the facts, we may rightfully picture to ourselves these two older portions of Oregon, three hundred miles apart, with the Pacific Ocean flowing freely between them.

The steady action of the surf upon the shore line would soon crumble the softer portions of these newly elevated materials until a wave-washed sloping beach was made a progressive margin for each region. It was on these sloping beaches that the life struggle first commenced, and on comparing again the fossil forms from each, the kinship of the two regions becomes at once plain.

Twenty years ago, Mr. Day, of eastern Oregon, found a small group of fossil shells on the elevated plateau between Canyon City and Prineville. Two of these shells, one of them a Pecten, the other a Pholadomya, were finely preserved and soon found their place in history. In time, Shasta County, in northern California, was found to have a like exposure of these Jurassic shells. Still later, Mr. Huntington, of Baker County, threw a flood of light on the extent of the eastern region by obtaining another group of fossil shells which proved to be well defined Jurassic, found twenty-five miles north of Burns. This searchlight was soon followed by another. A friendly geological student found in northern California a Triassic shell, the Halobia, recognized as of determining importance, while the same season a surveyor brought more of these shells from the region of Wallowa Lake,* plainly Halobia too. But a still earlier chapter of the Shoshone region has been revealed by a group of Productus shells in the Carboniferous rocks of Baker County and Rhynchonella shells collected on Beaver Creek in Crook County, also of the Paleozoic age; while in the Siskiyou region, we find fossil corals in Carboniferous limestones of Shasta County and Paleozoic fossils from southern Oregon.** These facts readily suggest a nucleus for each region during the Paleozoic age, an extension of these in the Triassic and Jurassic, and a continuous development of both regions throughout the Cretaceous. For the Trigonia, a bivalve shell of Cretaceous times is abundant in both regions, and of the three

*The Gold Belt of the Blue Mountains of Oregon, by Waldemar Lindgren, U. S. Geol. Surv.

**Note. Lindgren tells us there is also a small area of Gneiss of Archean age west of Sumpter in Baker County. This rare spot of Archean takes us back to a very ancient period in the earth's geological history, even to the dawn of plant and animal life.

Ammonites. *Plate IV.*

or four species most abundant on the coast, all are common in the fossil beaches of both lands. Two species of Trigonia may be seen on Plate II.

A handsome marine gastropod allied in form to Actæonella and Actæonina of this same period, and closely resembling the land shell Bulimus, is found abundant in the rock of both regions.

But the group of marine fossils that attracts most and holds the collector's attention longest is that of the chambered Cephalopods. Of this fine group the highest in rank are the Ammonites, a division of Cephalopods that was destined to become entirely extinct with the close of this Cretaceous period. The Ammonites were marvelously abundant along these beaches, and in both regions small inferior forms closed the history of their race.

In the earlier periods of their history, these lands present themselves to the geologist as similar in age, in structure and in their relations to the ocean; for throughout the whole time a like environment enwrapped both regions. A notable separation occurred to them with the close of the Cretaceous period. A geographical barrier was destined to separate them which next calls for our attention.

Physical geography has scarcely within its domain a more interesting group of facts than that which goes to make up our conception of a mountain range. We have here the necessity thrust upon us of imagining an original, nearly level ocean bed lifted to the form of a great upfold of the crust of the earth to a height of fifteen or twenty thousand feet and extending across continents. Such an upfold as that includes the Alps, the Caucasus and the Himalaya Mountains in one belt of a continent; or the Andes, the Mexican Cordilleras and the Rocky Mountains in another.

During long ages of quiet life on our lands of the Pacific, the forces of the natural world commenced the work of building another of these world girdles; one whose lines of progress were destined to pass between these lands of Shoshone and Siskiyou, and cast into separate and dissimilar currents their subsequent history.

A colossal sea-dike was slowly rising from the bed of the ocean, extending from what we now call Lower California, through what is now Oregon and Washington to the Aleutian Islands, a mere sea-dike for a long time, only a barrier between contiguous waters; then through other ages a ridge of elevated hills; then later one of the world's mountain wonders, the Cascade and Sierra Nevada Range. This new feature in the geography of our western coast was destined to become the great shaping force in its subsequent geology.

For so great a working agency, because of its relation to our narrative and the necessity of frequent reference to it, we need a distinctive name and shall call it the Cascade Barrier. In its line of progress it passed between the lands of our story and thenceforth Siskiyou and Shoshone became progressively different in environment.

Before the elevation of the Cascade Barrier the two lands of Siskiyou and Shoshone were dominant features of the region, and the geography was simple. The later multiplication of mountain masses removed this feature of simplicity and yet, in each case the older land became a mountain mass towering above extending plains, and retaining its dominance as a feature of the landscape. And further, as the later environment of the one was entirely marine, and of the other entirely fresh water, it will be most convenient to treat the ongrowth of the Siskiyou region as including all west of the Cascade Barrier; and all the region east of the barrier as the successive development of the Shoshone, designating one as the Shoshone Belt, the other as the Siskiyou Belt.

CHAPTER IV

THE LAND OF SISKIYOU

It will be remembered it was stated that the elevation of the Siskiyou region was accompanied with so much of violent force, and consequent heat, that the limestones of the mass were turned to marble, the argillaceous sediments into slate, and between and through these, beds of granite and other eruptive rocks were forced into the positions they now occupy in the mass.

It will be borne in mind, too, that the uplifted nucleus of the land was crowded to the surface with so much of heat and pressure as to destroy the outlines of many of the organic forms it contained. The land once elevated, the resistance to this crushing force ceased, and a long period of quiet succeeded.

It is just here, at the dawn of a new chapter in the life record, that the geological history of this Siskiyou region opens to the collector of fossils its richest records. Slowly the drifting currents of the ocean brought to its shores the eggs and seeds of sea and land, to open there into life under new conditions of environment. The beaches upon which these lodged were rough and inhospitable. The sea shells that just struggled for life, are found

Aralia Leaf. **Plate V.**

strangely cradled among rough, sharp pebbles, for soil and mud were scarce as yet.

The mass of conglomerate rock photographed on Plate III., finely describes this opening stage of a new geological period. It is from the base of the newer deposits, and the pebbles it contains are fragments of the older formation. These were broken off from its cliffs, worn partially smooth by the tossings of the surf, mixed with fragments of shells and then cemented into this conglomerate. The shell most abundant along this ancient coast line was a Trigonia, one entire valve of which may be seen in our illustration, surrounded by hundreds of fragments of others broken by the surf, then covered by scanty sediment, and so assigned a permanent place in this rock. These shells were multiplied, and others were rapidly added, until marine life filled every available nook of the whole coast line. The geological period whose beginning is thus marked was the Cretaceous. This is to say, the period in which these shells lived was that in which the chalk of France and England was deposited.

In other words: At the time when the great body of Europe was yet under the deep sea that was slowly depositing in its profoundest depths this chalk, the far off Siskiyou region was bordered by this ancient sea beach, on which these shells struggled for life. Later deposits on this beach as seen in layers a few feet higher and therefore more recent, show a notable change of material as well as fossil contents. The newer life of plant and animal added fresh material to the shore line deposit, and as a result, the smaller particles of the rock show increasingly abundant other life than that of the coarser shells. Sheltered bays lent their protection to fragile forms of life, and these so multiplied that the whole mass of the rock becomes eloquent in its record, for then as now, marine life most abounded in such sheltered places of the beach.

The chambered shells represented by the Nautilus and the Ammonite are found widely distributed in the rocks of the Siskiyou region. The Nautilus survived all the changes of the period and passed on to later times along the newly formed western coast; while its cousin, the Ammonite, of more complex internal structure, became extinct with the close of the Cretaceous. Three of this Ammonite family from the Siskiyou region are shown on Plate IV. The causes which brought about its extinction seem to have been connected with the complexity of its structure rather than its environment, for it drops out of geological history all over the world at the close of the Cretaceous period.

Bivalves of the Venus and Mactra families were also numerous, and smaller shells in increasing abundance filled those old beaches with beauty and wealth of scientific materials.

THE LAND OF SISKIYOU

In our narrative of the greater geological changes along the Pacific Coast, we found the gathering forces of a great geological revolution were taking form and direction, destined to change the whole western face of the continent. The colossal sea-dike referred to in the previous chapter as the Cascade Barrier was slowly rising from the bed of the ocean, its line of uplift falling a little eastward of the Siskiyou region, and ultimately resulted in the permanent separation of the two oldest sections of our state, for the sea coast character of the Shoshone was gradually changed to an inland fresh water region, while the Siskiyou continued its marine environment.

This change having occured at the close of the Cretaceous, the continuation of the history would open upon the Eocene period. The geological sediments of these Eocene times west of the Cascade Barrier were those of shoaling waters, chiefly sand-stones and often abounding in finely preserved marine fossils. An extensive belt of this rock not covered by any later deposit may be seen reaching from Denmark, a few miles north of Cape Blanco, to a point near the mouth of the Coquille River, reappearing again south of the entrance to Coos Bay, where fine exposures of fossils are found. Among these are well preserved Cardita planicosta, that may be seen near the lighthouse at Cape Arago, in rock tilted to an angle of sixty degrees. The rock reappears at the mouth of the Umpqua and makes the line of picturesque bluffs along both banks of that river to Scottsburg and is again extensively exposed along Elk Creek to Drain.

Further north, the Eocene reappears around Philomath and Corvallis. And here again one realizes the great thickness of those Eocene beds; for in spite of the evidence these hills furnish of long continued deep weather-wearing they present today a feature of striking magnitude. In several places among these Benton County hills, Cardita planicosta attest the Eocene age of these rocks.

Another exposure of the fossils of this great belt of Eocene is found near Albany. A low range of hills tends eastward from a point north of Corvallis, compelling the Willamette to run eastward for ten miles or more; and near Albany containing our characteristic Eocene shells in fine condition and goodly numbers. This range of hills between Corvallis and Albany is a mere off-shoot of the main belt which continues on unbroken to the Columbia River.

At Astoria, the Columbia River cuts into this Eocene belt and exposes another Eocene fossil,—the chambered shell Aturia*—a beautiful fossil and important because it proves the rock containing it to be Eocene.

*Note. The term Oligocene has been applied to many formations which were formerly classed as either Upper Eocene or Lower Miocene. The Aturia beds of Astoria are now designated as Oligocene by Dr. Dall of the Smithsonian.

The magnitude of these Eocene deposits, estimated in the light of what one sees along the coast line from Cape Blanco to the mouth of the Umpqua, strikes one as very great; while the uniformity of their materials and the abundance of their fossil remains alike, speak of long continued uniform history.

Anyone may verify these convictions by examining the thickness, uniformity and abundant life of these rocks on the North Umpqua, twenty miles east of Roseburg; the thickness of deposits along both banks of the Umpqua and Elk Creek from Elkton to the ocean; and at Cape Arago where again they are richly fossiliferous.

Let it then be accepted that this whole region enjoyed a long period of comparative quiet after the Cascade revolution. It was during this period that the greater part of the coal deposit of our Pacific Coast was formed. The coal producing period in the United States east of the Mississippi River was the Carboniferous. The Rocky Mountain region looks to the Cretaceous for its supply of coal, while we, of the Pacific Coast have the later coal of the Eocene.

The most important coal deposit so far discovered in Oregon, was formed on the northern shore of our Siskiyou region. Doctor Diller, who has made a careful study of this field, reports the land as gradually sinking during the Eocene period, although with long intervals of rest. At times the field would be covered by fresh water or brackish swamps on which flourished a rank growth of vegetation destined to become a seam of glistening coal. Then for a time the sea would gain upon the land, leaving a deposit of sand and mud which later formed a layer of sandstone and shale, containing Eocene shells. Then another interval of rest and filling in of sediment would produce the broad expanse of rich swamp vegetation, which in time contributed its seam of coal. This same process repeated again and again through a long period of time gave us our Coos Bay coal. Later the old level lines of deposit became tilted and broken into endless confusion of outline.

During the Eocene, the Olympic Mountains of Washington* formed a large off shore island, and between that island and the Cascades were many coal swamps reaching their long arms into the low-lying Cascade Mountains. Here in the Puget Sound basin, all the coal producing conditions of Coos Bay were repeated, and existed for so long a time that two vertical miles of coal seams and stony sediments were laid down during the Eocene. Some of the Washington coal was made in mountain lakes; shallow, marshy lakes, whose beds were slowly sinking and receiving the wash and wear of the surrounding mountains.

But the completion of the plan of North America calls for another world belt, and it comes this

*Note. Washington Geological Survey, Vol. II.

time as another upfold of the crust of the earth beginning again under the sea bed and reaching from California to Alaska. Its line falls upon our land of Siskiyou, and continues northward and southward—a sea-dike in its early stages, later a range of hills, and still later a range of mountains—the Coast Range of the future Pacific Coast. It will at once strike the reader that this new event has in it several points of likeness, both in itself and in its results, to the elevation of the Cascade Barrier already described. This conclusion is entirely just, for both of these upcrumplings became permanent ranges of mountains; both cut off permanently portions of the Pacific Ocean, changing these to inland belts of water, later to drainage troughs, and ultimately into valleys; while the Shoshone became a dominant feature of the newer land east of the Cascade Barrier, and the region between the Cascades and Coast Range was dominated by the land of Siskiyou.

Inasmuch as the upfold along the coast line is apparently the final one in the structure of North America, and as two others of like structural character preceded this, a brief consideration of the apparent causes that underlie these surface movements may not be out of place.

The first of the series within the range of our subject was that which built up the Rocky Mountains; the second that which forced up the Cascade Mountains; the third, this latest one, the elevation of the Coast Range. In all these movements, the work was begun in the crumpling up of the bed of the sea. In proof of this one has only to point out the fact common to all three regions; that the sea mud, now hardened into rock and elevated to the shoulders of these mountains and constituting the bulk of the mountain mass, abounds in sea shells, many of which lie apparently undisturbed in the very mud banks in which they lived. Another fact common to all three regions is the evidence they furnish of the great force and heat that accompanied their elevation. This is seen in such portions of the masses as offers most resistance to this elevating force, changing in such places as by fire, the very structure of the mass. These facts and the great extent over which they prevail seem to prove a cause at least as extensive as the miles they cover.

We have now opened before us a new chapter in the geological history of our coast. The framework of the new order of things is traced in bold features. Its greater features are twofold; first, a mountain range lifted from the bed of the ocean nearly a hundred miles westward of the Cascade Barrier; second, a related trough between these two extending along the entire length of the coast from southern California to the Alaskan Peninsula.

Throughout this extent the unity of the elevated mass receives recognition under the name of the Coast Range. The unity of the related trough

Skull of Fossil Seal—Desmatophoca Oregonensis. *Plate VI.*

has not received a name to cover its whole length; yet a moment's inspection will show the same unity among the parts of the trough that is so carefully noted in the uplifts. The San Joaquin and Sacramento Valleys of California, the Willamette Valley of Oregon, the trough of Puget Sound, through Queen Charlotte Sound—all these are but parts of the one continuous trough that separates the Cascade Range on the east from the Coast Range on the west.

The incompleteness of this new coast barrier was such that the ocean water had for a long time free access to the Oregon part of the depression, now the Willamette Valley.

The unity of movement, shown in the lines of continuance of the mountain range and the trough between this and the Cascade Range, presents several breaks; but these broken links of the chain are explained by the fact that the imperfection is always where the new line of upheaval crosses an older upfold of the crust. In proof of this, notice that the Californian part of the trough is interrupted by the older uplifts over against the Siskiyou region; that the trough which begins again in southern Oregon is interrupted by cross lines of earlier upfolds, disturbing its development till it crosses the Calapooia Mountains, after which, the trough, as the Willamette Valley, regains its normal relations to the wider outline of the continent and passes on as the

Cowlitz Valley. Here at the head of this valley occurs another short break in an old Eocene mass and then the trough falls into line again in the depression of Puget Sound. A fainter continuance of this trough may be traced northward through Queen Charlotte Sound and even farther on.

The geographical extent of these changes, together with an approximate parallelism of the working forces that produced them, would seem to indicate a world-wide cause, modified along the Pacific slope by the coast conditions, but unquestionably originating from a continued shrinkage of the crust of the earth. The earliest yielding to the force produced the Rocky Mountains, the next the Cascades and the latest, the Coast Range. These were so many upfolds of the shrinking crust, each with its correlated troughs or down-folds resembling the shrinkage on the surface of a last year's apple.

This latest of these continental upfolds, the Coast Range, having taken its place among world facts, an outline of its geological relations will be important.

At the opening of the Miocene period, the coast upfold was only a line of islands, the geological materials of which were the recently elevated belts of Eocene. Between these islands were open passages through which ebbed and flowed the ocean tides, while the enclosed belt of water between the islands and the Cascade Range was wide and deep, a body of water similar to Queen Charlotte Sound of today. It was on the inner slopes of this sound that the marine waters of the period deposited their rich and varied record of the life of the times.

In this way the outline of our modern Willamette Valley was first blocked out, the framework of which still remains. Its rim has varied in altitude, its bed has varied in depth, but its original outline abides. The Coast Range of mountains was pierced by straits through which the ocean freely flowed till it washed the slopes of the Cascade hills. This free inflow of the ocean kept the enclosed body of water salt enough to promote the health and abundance of its sea shells, and no region of Miocene times has kept a fuller record of its life. And if now we are led to ask where the Miocene agencies found the materials for these later upbuildings of hills, we may remember that the great upfold we now call the Coast Range, once in place, settled into two sorts of materials; a denser, heavier portion that formed the mud sills, and a lighter topping lifted skyward. It was the work of Miocene weatherwear to form the skyward portion into graceful drapery for our coming valley. In short, Miocene winds and rain transported these looser Miocene materials into gentle inland slopes to be in time covered by the tides of the ocean and stocked with sea shells until it become a grand Miocene aquarium.

The progress of this Miocene work may be traced in the deposits of sediments along the inner slopes of this basin, the changed remains of which now border the Willamette Valley as thick beds of rock, aggregating in some localities hundreds of feet in vertical thickness. In many places they are marked by abundance of marine shells in the very sands and mud flats in which they lived, but all now changed to rock. Time has worn awav the less dense surface layers of this varied deposit, but the denser lower layers have in different degrees resisted this wear and formed the foothills in nearly the entire circuit of the present valley from the Columbia River to the Calapooia Mountains.

A fine exposure of these Miocene sediments may be seen in the excavated streets of Astoria, resting on a like exposure of the underlying Eocene—both groups of fossils, finely preserved, are often enclosed in concretions, which apparently formed around the decaying shells in quiet, deep waters. Sometimes the enclosed fossil is a crab, and in the shales of the underlying Eocene is a chambered shell, a near cousin of the Nautilus, the Aturia, of fine form and well preserved. Another interesting exposure of these Miocene fossils may be observed at Westport, a few miles above Astoria. The fossils here will be found at the base of the hill, south of the landing, and closely repeat the forms at Astoria.

A Heap of Bones. *Plate VI*

Still farther up the river at the base of the foothills back of Scappoose, these Miocene fossils are again seen, and here they underlie a good deal of volcanic rock, but on crossing over the mountains into Tualatin Plains, the Miocene shells are again found in their native sandstone. Southwest of Forest Grove where Coast Mountain streams cut through the Miocene foothills, the fossils are again very abundant and very fine—all Miocene. South of this, bordering Wapato Lake, they appear again in shales; and this border land of the Miocene may be traced by its fossils to the head of the valley at Eugene, where they appear in great beauty and abundance.

Passing northward on the east or Cascade side of the valley one strikes these Miocene beds at Willoughby's, West Point, Peterson's Butte, Knox's Butte, in the bed of the Santiam, east of Knox's Butte and so on till we reach Salem, whose hills are all fossiliferous.

The later geological record of the Willamette Valley is such as to suggest an explanation of its facts. The theory is that throughout the next geological period, the Pliocene, the whole region remained above water and so left no record, for if it or any considerable part of it was covered by water during Pliocene times, the camel, the horse and other mammals would be found fossil here. But later in the Champlain period it shared in the general land subsidence that then marked the whole

North Temperate Zone. The Willamette Valley, the Yakima Valley and the Walla Walla region were all again covered with water to the depth of several hundred feet. It is to the western Oregon share of these elevated waters that the name Willamette Sound has recently been given.

CHAPTER V.

THE LAND OF SHOSHONE

Thus far our attention has mainly been occupied with the western or Siskiyou region; we have now reached a point in our narrative at which the land of Shoshone calls for some notice of its specific features.

Its outlines like those of the Siskiyou of the Cretaceous period, are in part suggested by the backlying ridges of the changed older rocks, and partly by the uplying surface of the Cretaceous shore line. One sees a fine exposure of this Cretaceous at old Camp Drake in the Crooked River Valley, which plainly was once part of the shore line of the western spur of Shoshone. A like stretch of the northern shore line of the same spur may be seen in several localities in the John Day Valley—one of these on Rock Creek, and Spanish Gulch, east of the old Camp Watson; another farther west on the upper waters of Bridge Creek—the marine shells from all these localities being abundant, very fine and all Cretaceous.

Mention was made of backlying ridges of the older rocks that escaped the violent changes of the early history of both regions. A good example of this may be seen on an elevated ridge west of

Canyon City, the Jurassic rocks of which are plainly seen with their characteristic fossils unchanged.

But the grand work of building at this time was Cretaceous work, and the land of Shoshone went on enlarging its outline in times of elevation of surface and lessening its area in times of subsidence throughout the whole period of Cretaceous time. But at the close of this period when the Cascade Barrier had become sufficiently elevated to exclude the ocean from the eastern region, the inland waters slowly changed to brackish, then to fresh water. But these inland waters were very extensive. They were a part of the great watershed of the western slope of the recently elevated Rocky Mountain Range.

The discharge oceanward of this long watershed slowly divided into three areas of drainage;—the southern one laying the foundation of the drainage of the Colorado River, the northern one laying the foundation of the drainage of the Columbia River, while the middle region awaited the slower process of drainage by evaporation.

The result of these elevations of surface and erosion of drainage currents, slowly subdivided the enclosed waters until the interior of Oregon, to which region our story is restricted, became covered with extensive lakes connected by drainage links from lake to lake—till by progressive drainage the links grew into a continuous stream—the Columbia River.

Oreodon Type Head. *Plate VI*

It is with this lake period of the geological history of Oregon, that our narrative seeks next to deal, and it is surrounded by these conditions that our land of Shoshone develops the rest of its history. As these lakes became smaller and shallower the land extended, and these extensions so varied its stretches of hill and plain as to make it a region of exceptional wealth of beauty and variety.

That the climate was the climate of the rhinoceros and the palm tree is proved by the fact that the bones of the rhinoceros and the leaves of the palm tree are now found in the rocky sediments of its lakes. Sometimes, too, one finds the fruit and leaves of an alder, then those of a maple or an elm; often the leaf or branchlets of a yew-like tree closely resembling the California Redwood. In fact the leaves of the forest trees are large and of wide range of species, impressing one with the conviction that they are the product of a moist as well as a warm atmosphere. One readily sees from other than the evidence of its fossil leaves why this must have been the character of its climate, for in the early Tertiary times, the Cascade range of hills had not yet reached an altitude sufficient to exclude the warm, moist air of the Pacific Ocean from the land of Shoshone as the present Cascade Range does now in that same region. So in those days the moist atmosphere of the Pacific Ocean wrapped the land in its cloudy folds and shed upon its slopes frequent and refreshing rains.

But more than this: much of the present region of Alaska remained yet under the ocean during the early Tertiary times, and thus presented an open passage for the Japan current flowing north of these Oregon lands on its way towards Hudson Bay and the coast of Greenland; this cut off all accumulations of snow and ice between Oregon and the Arctic Ocean.

That great revolutionary movement, the elevation of the Cascade Mountains, while important in its relations to both, did far more to give variety of surroundings to the Shoshone than to the Siskiyou region. To both regions the changes it caused make the event itself the geological epoch of the times. In our narrative it will furnish us with a convenient division line between the records of the Cretaceous and those of the Tertiary rocks, for it was the closing work of Cretaceous times to separate, and the opening work of the Tertiary to begin the new record. But another feature of the region we are describing owes its existence to the elevation of this range. It is the enormous masses of volcanic rock that exist along the eastern slopes of the Cascades. Ordinarily volcanoes throw out well defined streams of lava, so that their masses are easily measured; but this upthrust of Cascade lavas was most of it in sheeted masses, bursting to the surface like sheets of water through broken ice, and freezing over its rents.

These broader sheets of lava seem to have been thrown out in the earlier stages of the period of eruption and to have left their traces along the eastern slopes of the range, while the letter narrower and shorter outflows built up more rapidly westward of the axis of the upfold. I recall an impression I once received of the enormous thickness of some of these lava deposits on the Des Chutes River, east of the Cascade Range.

At about thirty miles from The Dalles, the old Canyon City road crossed the Des Chutes. In its descent from the Des Chutes hills the road was cut into the face of the hill, giving a good exposure of its rocky materials. It was thus seen that in the upper part of the hill, loose materials, held in place by an occassional dense lava flow, characterized the road through a descent of sixteen or eighteen hundred feet. The remaining part of the road was simply a climbing down across the escarpments of twenty-seven to thirty well marked flows of lava ranging in thickness from twenty to fifty feet and aggregating not less than nine hundred feet. Through all of these, the Des Chutes had worn its channel without reaching the bottom of this underlying hard basalt.

Manifestly the upper and newer portions of the hill were built of lighter materials mixed more and more, as one ascended, with layers of volcanic ashes; and where the layers were of denser materials showing weathering and erosion between the eruptions to which they were due; at once suggesting that these eruptions were separated by considerable intervals of time.

In marked contrast with this condition of these upper layers seemed that of the denser ones of the lower division, indicating rapid successions of outflows with no time for weathering between them. The enormous eruptive activity indicated by these lower deposits must have occurred in early Tertiary times, and in sight of our land of Shoshone.

While these denser layers of basalt were poured out over the surface in the earlier stages of this great eruption, another feature marked its later stages —that of vast showers of volcanic ashes alternating with the denser flows of lava. An example of this form of volcanic agency is given from a modern instance of its results. What it lacks in magnitude it will make up in clearness of detail. The one offered is from notes of a journey across the Cascades along the line of the Middle Fork of the Willamette River.

At eight or ten miles from the summit, going eastward, the whole surface soil seems changed by a covering of yellowish dust, growing thicker as one approaches the summit. On the eastern descent the depth of this surface addition is greatly increased and its true character is plainly seen to be volcanic ashes. While carefully exploring for evidence of its thickness at its deepest, I became satisfied that it could hardly be less than twenty-five or thirty feet. It was distributed evenly as a fresh snow in still weather. Where it was at its greatest depth, the whole country was covered with a growth of forest trees and all of the trees were growing naturally from the surface of this ash bank. The trees of largest size were Yellow Pine (Pinus Ponderosa), some of them three or four feet in diameter and in good healthy condition. All this forest growth was on the surface of a sheet of volcanic ashes.

Evidently all this heavy timber came from a replanting that occurred long after the volcanic eruption. For aught that appears there, several successive generations of trees may have come and gone since that shower of ashes fell. This deposit of ashes lay lightly for a time; but rains and snows, falling on it, settled it down with the help of its own gravity. Where it is now twenty feet deep, it must have at first stood expanded to forty or fifty feet, a depth inconsistent with the probability of survival of any forest tree on which it fell. It slowly diminishes as one recedes from the mountain until at forty of fifty miles it disappears.

After a good deal of search for some exposure of the old surface upon which all this spent fire material had fallen, I was fortunate enough to find one. At a place undermined by the Des Chutes River, I found a fine section of the bank with about ten feet of the condensed volcanic ash covering seven feet of dark rich soil with traces of other times.

This recent instance of the eruption of volcanic ashes furnishes an impressive lesson in geology, showing as it does the extent to which such volcanic products were capable of modifying the sediments of the waters of this region during early Tertiary times.

But this work of covering the hillsides with showers of soil capable of sustaining forests of pine, was not the only notable result by which these showers of ashes may be traced. The prevailing winds were westerly, and the drift of ashes to the eastward; and this being a country of many lakes in the times we are describing, much of this volcanic ash would fall on these lake surfaces and become an evenly distributed sediment, covering up and so preserving the remains of bird and beast in stratified volcanic ashes. A very interesting example of this kind one finds near Silver Lake, eastward of Klamath Marsh, where finely preserved fossils are so found.

These two classes of products of volcanic activity have worked together—the volcanic flood and the volcanic cloud, each doing its part in fashioning the future of the land of Shoshone.

Yet another feature of great importance in modifying the future of the Shoshone region, remains to be described—the enormous accumulations of lake sediments. All the wash and wear and drift of the land was of necessity towards the lowest places —the lake beds of the interior, and the sea bed of the coast. The dust of the atmosphere, the sands of the river current, the mud-wash of the storm, the wear of the coast line—all these ultimately drift toward the deepest places, and because these deepest places have but little current—and only the finest of the muddy sediments go into these currentless depths—it results that deep oozy mud covers the bottom of such places and accumulates there. That the mass of this ooze keeps increasing with time, needs no argument; and that the increasing weight of continued accumulation slowly presses the lower portion into various stages of solidity, is only a question of time and gravitation.

Such a natural process of growth of lake sediments must have occurred in the deep lakes of eastern Oregon in those early Tertiary times. Can we find and verify them in that region today?

At a point sixty miles south of the Columbia River and about forty miles east of the Des Chutes, on the old road to Canyon City, the road gradually ascends from Antelope Valley to Cold Camp, and from Cold Camp to Currant Creek Hill, from whose ridge looking eastward a wonderful panorama opens to one's view. It is the region known to the old miners as the Potato Hills. Our Currant Creek Hill is simply the elevated shoreline of an old lake fifteen hundred to eighteen hundred feet above the once muddy masses that formed its bed. From this Currant Creek Hill, we are looking eastward into the very depths of that old lake bed that once swept around the western spur of our land of Shoshone.

And now, if after an hour's enjoyment of this grand geological landscape, and carefully noting the varied shades of the picture, we examine more minutely the surface features, we may easily observe three or four drainage depressions through which as many creeks have carved their beds, and through which they now empty northward into the John Day River. The nearest of these, and almost at the foot of the descent before us, is the depression of Currant Creek. Beyond this, a like irregular depression almost concealed by the hills it has carved, lies the bed of Cherry Creek.

All this is simply a vast system of carving done by these streams in the muddy sediments of one of these old lake beds, and now turned up to the light of study.

The black stubs of inky shade are so many dykes of eruptive rock burst through the mass. One of these, a dyke of columnar basalt, is exceptionally instructive. Seen at the distance of two or three miles it seems a vast wood-pile—a giant's wood-pile. Seen at closer range it is found to be a stack of rock prisms of great regularity, evidently once forced up through this old lake sediment, from which atmospheric wearing has long since cleared

Oreodon, Broad Triangular Crown. *Plate IX.*

away the softer covering, leaving our wood-pile a picture itself.

These eruptive rocks, in their relation to the general landscape, seem like ink blotches on the face of a painting. If we look carefully at the other rocks we shall find them all hardened sediments—fresh water sediments—for whenever they contain shells these are fresh water, or land shells, but never marine.

The aggregate thickness of these sedimentary rocks of Tertiary times, between the Shoshone and the Cascade Mountains is very great; not less than three thousand feet. If this seems an extraordinary depth for a group of lakes, one need only point out the almost certainty that throughout the whole time of this vast sedimentation the Cascade Mountains were subject to further uplifts, and the depressions east and west of this line of upfold were subject to correlated subsidence, so that the earlier sediments of the Willamette Valley on the west, and the beds of these interior lakes on the eastward, by their gradual sinking, increase the capacity of these depressions for continued sediments. This borne in mind, the depth of even three thousand feet will not seem startling.

Throughout this whole mass of lake sediments, fine and dense where it was deposited in very deep and quiet water, fine sandstone where the inflowing streams left the traces of their emptying—all occasionally interstratified with beds of volcanic ashes—are interspersed the leaves of plants, broken fragments of wood, pieces of bone of fish, of fowl or of mammal.

Now all of these leaves, fragments of wood and pieces of bone, are enough different from the leaves and bones of today to prove them belonging to other times, for they tell the story of other forms of life than ours. It is the great number and variety of these fossil remains of former life, interspersed throughout these three thousand feet of sedimentary rock, that gives this old lake bed its marvelous interest.

One is liable at any turn in a ravine to find protruding from its ledges a fragment of a skull, a bone or a tooth, ranging from those of a squirrel to those of a horse. But all these under ground records are remains of things that once abode above ground, and so really describe the former inhabitants of the Shoshone and its environment of lakes.

If now from our point of outlook on the crest of Currant Creek Hill, looking eastward, we may imagine the waters of that lake returned until they submerged the whole of the John Day Basin, we might see across these returned waters fifty miles away the ancient Western spur of the Shoshone land of the early Tertiaries. True, it today lacks the palm tree on the shore and the rhinoceros near its waters. But their graves are here, monuments that speak eloquently of other times.

While these sediments of mud or sand or ash all come from the wear and wash of the land transported into the lakes by winds or streams, there is another sedimentary deposit indigenous to the lake beds themselves, and so different from the others as to call for a word of special mention. It it known as Diatomaceous earth. It is found in the form of a stratified, white or yellowish white rock, often so light as to float on water and sometimes mistaken for chalk. It is often sold in the shops under different names as polishing powder, and as such is often named Tripolite. On examination under a good microscope it is found to be made up of the remains of one-celled plants whose cell coating is silica, and it is the accumulation of these shells, or coatings that makes the mass of the rock.

Recent deposits of this material are found in both sea and fresh water, sometimes in swamps and in peat bogs. Extensive beds of it are found fossil along the upper Des Chutes River in old lake bed deposits. A fine bed of it may be found in a sloping bank of Three Mile Creek, two or three miles South of The Dalles.

Geologically older than these deposits just mentioned, is a bed of this rock found in the upper John Day Valley, underlying the Pliocene rocks of the region and tilted at quite an angle, while the Pliocene is nearly horizontal, a fact that would at once place it among the earlier Tertiaries.

An interesting outcrop of this may be seen at the old Belshaw Ranch on the Canyon City road in the John Day Valley. The rock itself floats on water, is as white as chalk, and contains, besides its microscopic plant cells some finely preserved leaf prints of oak, maple and other forest trees. Like chalk, this rock is made up of the dead shells of a one-celled organism, but the chalk is the cell covering of an animal; the Tripolite is the cell covering of a plant. And further, the chalk is calcareous material, the Tripolite is always nearly pure silica. Now this Siliceous Diatom earth was very widely diffused in the waters of the Shoshone region and so added its share to the various sediments that form its rocks of today.

Such were the materials that entered into the composition of the land of Shoshone, and such are the materials that constitute, in a more or less altered condition, its yet untranslated history.

THE LAND OF SHOSHONE IN ITS WIDER RELATIONS.

GEOGRAPHICAL.

Thus far our narrative has had to do with occurrences apparently local and dissociated from facts and events that shaped the history of the rest of the world. Our story now needs to take on its relations to this wider circle of changes, the geographical progress of other regions.

As the Cascade barrier was slowly changed from a mere water dyke to a range of hills, and still later

Orcodon, Long Nasal Bone. *Plate X.*

to a vast range of mountains, increased elevation lifted it into an atmospheric agency quite as important as its previous marine one, for when it reached the altitude of a mountain range it excluded the moist, warm current of the Pacific Ocean and thus surrendered the interior to the dry, cold winds of the continent eastward.

Yet another of these barrier functions remains to be ascribed to the Cascade Range. Its uplift along the coast of Alaska made it a barrier to the flow eastward of the Japan Current of the ocean. The present extended plains from Alaska to Baffin's Bay would warrant the conclusion that before the elevation of the Cascade Barrier at Alaska, the Japan Current must have flowed over those stretches of low country on its way northward. The effect of this, as previously noted, would be to prevent accumulations of snow and ice in that region, a condition of things which would be very effective in modifying the climate of the land we are describing. Yet such an inflow of a vast tropical river from the ocean itself, must have existed till turned aside by the upfold of this Cascade Barrier along the coast of Alaska.

But there is a still wider view of its world relations than this one of the Pacific slope; for while this Cascade Barrier was making a geographical separation between our lands of the Pacific Coast there was an extension of the Gulf of Mexico northward into what is now British America, covering much of the region now occupied by the Rocky Mountains. The same crumpling process that elevated the Cascade Barrier, by a like process of elevation closed this American Mediterranean to the ocean, and also added to the height and breadth of the already begun upfold of the Rocky Mountains. This change was closely followed by the conversion of the inclosed waters of the region from salt, through brackish, to fresh waters.

And yet a still wider relationship may be mentioned; up to the time when the Cascade Barrier was separating our lands of the Pacific, Western Europe, from the British Islands to the Black Sea, was covered by a deep ocean over whose bed had been slowly deposited the cast off calcareous shells of a Protozoan animal, the Globigerina. This accumulation of life remains, hundreds of feet in thickness and extending over a length of six hundred miles, was brought to a close by the elevation of the sea bed, its calcareous sediment to be known in after time as the chalk beds of Europe.

Now this shrinking and the resulting crumpling of the surface seen in this light, becomes a world fact; its manifestation in the Cascade Barrier, its other manifestation along the line of the Rocky Mountains, and the still further one in the elevation of the chalk beds of Europe, are but three links in the one chain of force. It is this European link that gives its name to the epoch, the Cretaceous (meaning chalk). And the close of this period was a time of great change, a revolution in the geological history, marking the passing away of the older and the introduction of the newer forms of animal life. To accomplish this result, the great types of animal life went through rapid changes. The dominant forms of vertebrate life of the Cretaceous period of land and sea, were reptilian; the dominant forms of the new period were mammalian.

In the light of these facts there is a striking fitness in the name geologists have given the period that follows the Cretaceous. They call it the Eocene—the dawn of the recent.

When the violence that accompanied the Cretaceous revolution passed away, quiet was restored, and life—land life—took its new tendency on our land of Shoshone.

CHAPTER VI.

INTRODUCTION TO THE LIFE OF THE LAKES

In our narrative of the changes that occured in the Siskiyou region, the record we tried to translate was made by the waters of the ocean; in carrying out a like inquiry for the Shoshone region, we have found ourselves shut in to the record of the lakes.

The ocean has no break in its record, for the ocean always has a beach and the beach always makes its record, whether kept or lost in later changes. The muddy sediment of the lake lasts with the lake, but ceases to write when the lake dries up or drains off, and its record closes. The lake, then, keeps the geological record of the interior of the continent only while the lake lasts, the ocean keeps that of its external margin and is continuous. It will be remembered when the great sea dyke that formed the foundation of the Cascade Mountains was first elevated above the surface of the sea, it at once cut off from the ocean and made an inland sea of the great body of water between that dyke on the west and the shore-line of the Wasatch, Bitter Root and Coeur d'Alene Mountains on the east. Near the northern por-

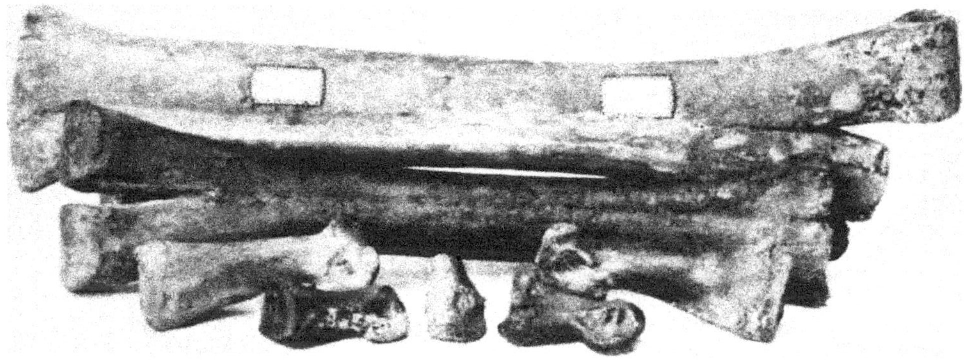

Comparison of Camel and Auchenia Bones. *Plate XI.*

tion of this inland sea, the large irregular body of land we have called Shoshone must have formed a fine relief feature with so vast a water waste. The level at which the water stood along its shores would, of course, determine its extent and outline; and these must have varied from age to age, for surface disturbances have left many records of their continued activity.

The slow elevation of this region developed, as already described, a drainage system southward that in time became the Colorado River, along whose water-shed can be found the record of its successive changes. The middle portion, with its grand list of Tertiary lakes, we may dismiss as outside our theme for consideration, but we will center our thought on the remaining third or northern portion of these waters.

After the close of the Cretaceous period, as the sea was excluded from the interior, we would thereafter expect to find only fossils associated with fresh waters, and these alone are found there. The contrast between these two kinds of sediment is striking. One of its many instances can be observed on Bridge Creek, in Wheeler County, near the town of Mitchell. Here one can see an extensive table land covered with a thin soil scarcely concealing the rock below it, and this rock abounding in sea shells; Trigonia, Actæonella and Ammonites, all plainly Createceous, while from the plateau on which they occur one may look to the westward a mile or two over a lower level filled with later sediments. These, one soon learns to distinguish as Miocene fresh water deposits of the Bridge Creek Oreodon beds. Here are brought into sharp contrast the old basement floor of the Cretaceous seas and the later deposits of the Tertiary lakes, both forming today the western slope of what was once our land of Shoshone. It is to be understood here that in all that is said of lake records, there is implied a sediment of mud containing casts and impression of the living things of other ages, of things as real as the writing on Chaldean pottery or hieroglyphics upon a piece of Egyptian papyrus. The sediments of these lakes are in effect, so many openings into the past—openings from which the curtains of time are drawn aside so we can see through them the vista of the ages.

Two of these lake beds are of special importance to us because they represent, distinctly, separate life records. The lower and older strata contain a marvelous wealth of fossil life running through a long stretch of geological history. The upper or later sediments are of more restricted surface, but of wider scope of life forms. The lower lake remains would represent the early and part of the middle Tertiary, while the upper would belong to later times. The oldest of these lakes may be traced along the lower reaches of the John Day River, excavated by the wash and wear of that stream. A southern extension of this sediment was laid bare by the drainage of the Crooked River, an eastern branch of the Des Chutes. Both of these will be described as lower lake deposits, for they belong to the same geological horizon.

The lowest of these deposits abound in well preserved leaf impressions, many of them equal to the finest engravings, the original carbon of the leaf furnishing the printer's ink. As fitly illustrating these fossils there is introduced on Plate V. the print of an Aralia, a cousin of our thorny panax. A fine impression of a fan palm was also found in these same rocks a few years since. The palm required a subtropical climate, and such without doubt was the climate of this region during the Eocene period.

These fossil remains of forest trees of the early Tertiary are so abundant and so finely preserved that one is disappointed at the entire absence of the bones of land animals from these early records, their fossils being so abundant here in the next age, the Miocene. But it must be remembered that at the close of the Cretaceous period the mammalian forms of life were but few, and these not widely distributed. A few small marsupials, and some of still lower type like the Duck-bill, were all the world had of mammalian forms at that time. Throughout the Cretaceous period, the life of the world's vertebrate animals was nearly all reptilian; the air, the land, the waters teemed with reptilian forms of life. Mammals of the two orders named had recently multiplied in North America as well as in Europe and Asia, but it was not until after the great mountain revolution of the Cascade Barrier, and therefore until the Eocene period, that the world received its stock of highly varied land animals, the mammalia. If, then, it was in the Eocene period that the orders of mammalian life were so varied and spread over the continent of North America, a glance at the geography of the period will help to determine the relation of this fact to our land of Shoshone.

All through the Cretaceous period, there existed where the Rocky Mountains now stand, a deep Mediterranean Sea. When at the close of this period, the Rocky Mountains were elevated, their mass was raised from the bed of this sea. As part of the result of this uplift, the ocean waters were thenceforth shut out from the interior of the continent, and the hollows that remained were filled with fresh water and became lakes. Now, into these lake beds on both sides of the newly elevated mountain range, were washed the skeletons of the animals of the Eocene period, and so preserved as fossils. These old lake beds are the Bad Lands of today. The fossils from one of these Eocene lakes, just east of the Wasatch Mountains, are extremely interesting and have thrown much light upon the early development of mammals. Just why the remains of these same Eocene mammals have not been found in Eastern Oregon is one of our unsolved problems. Either the Shoshone region was cut off from the Wasatch region by intervening waters, or if these animals lived in Oregon, their remains may yet be found.

The fact that fossils are found in lake sediments furnishes no proof that the animals to which they belong were water animals, for although many of them were such, a much larger number lived upon the land. This result comes in this way: we have given a large lake, hundreds of miles in extent, into which several rivers and very many smaller streams empty—all flowing from the surrounding hills. Now into these streams the wash and wear of the whole surface of the hills are slowly drifting to the lowest places. Once dumped into the flooded stream bed, all this is taken up by the current and washed down to the lake and out into its depths to be covered in its muddy sediments. This granted as the ordinary course of things, it is readily seen that there is no sick or wounded animal of the hills passing to the banks of the streams to drink, but is liable to die there and the next flood will wash what is left of its skeleton into the lake where it is soon buried in the soft mud of its sediment. In this way the skeletons, or scattered bones, of not only the larger animals, but those of rabbits, rats and squirrels from the hills are borne to their resting place in the depths of these waters.

These lakes of Eastern Oregon continued to receive into their waters, and through them into their sediments, the remains of forest and field, burying them up safely in their profoundest depths and thus securing them from decay. The perfection of this mode of preservation is well nigh complete. The minutest vein work of the leaf; the insect that fastened to its surface; the seed pod or capsule of the plant; the pores of the wood; the sutures of the animal's skull; the epiphyses of the bones, enabling one to tell at a glance whether these belong to the young or the old; the minutest lines of age or accident on a tooth, are all preserved with marvelous faithfulness to the life type of the period, so that no family of plants or of anmials is likely.to have lived on the borders of these lakes during that long stretch of time, covering hundreds of thousands of years, without sharing in the record of its fossil history. It is from such archives that materials have been collected for a geological record of Oregon's past.

In describing in such rapid succession the changed condition of this region and of the mountain barrier that walled it in on the west, there is danger of giving the impression of a continued record of stupendous violence. To correct this wrong impression, one needs to think of great length of time over which these records of changes stretch, as extending over millions of years; and of very many of these changes themselves as scarcely attracting attention while occurring. Its measure of result was the growth of long ages, if not eons. That there were times of great and sudden violence, there can be no doubt; but these were of short duration when compared with the long ages of quiet growth for plant and animal that intervened.

The reader will find Plate VII. engraved from a photograph of real fragments of the bones and teeth of different animals all gathered from the mud of this lower lake. A curious fact in regard to these ought to be added here, that the teeth figured on this plate are all changed to agate and the cavities of these bones and teeth are often filled with crystallized quartz. In Oregon the term Miocene will always apply to the rocks and fossils of this period of quartz-filled cavities. This fact of quartz infiltration is also observed in some belts of Miocene rock of the Willamette Valley in which the cavities of fossil clams contain perfect casts of chalcedony or agate. In both of these localities the liquifaction of the infiltrated silica seems due to the pressure and heat generated under the weight of later deposits of sediment, since removed by the wash and wear of atmospheric erosion. The silica infiltration occurred after the enclosure of the specimen by the sediment. The shell drifts into deep water and sinks into the soft oozy mud of the bottom. As long as there remains any crack or chink of the shell accessible to this mud the weight of the water presses in more. In those shells that are thus entirely filled the process of filling up the inner cast is completed. But if through any cause the shell is completely enveloped by the sediment before the inner cavity is filled, then this upper and unfilled portion remains empty. Now it is this remnant of space that is filled with silica. The result is striking, for we have the exact cast of the inner cavity of a bivalve, its lower half a dark opaque sea mud, its upper half a fine translucent chalcedony of rich carnelian tint. If the shell is so entirely closed as to wholly exclude the muddy sediment, then the whole inner cavity is lined with silica, clear as crystal, or if almost filled with quartz, as often happens, then a small central cavity only remains, and this often filled with vapor of water. As these cool the water settles to the bottom and an air bubble rises to the upper level, making the water agate so much prized on our coast.

Or if we take the same process as it may be traced in a different fossil, we have here the jaw of a rhinoceros. We have the completed result in this fossil, and the entire process is this: the jaw with its teeth in place is washed into the deep waters of an eastern Oregon lake; it sinks into the soft mud of the lake bottom. In time the mud is pressed into every accessible crevice of the bone and its dental contents, but, in the teeth that are entire, there remains a whole set of cavities inaccessible to this lake mud. It is into these nerve cavities that the liquid silica penetrates and in which it is preserved.

Oreodon, Size of Fox. *Plate XII.*

LIFE OF THE LAKES

The rocks made by the sediments of the later Oregon lakes are looser, lighter, more porous; the fossil leaves are more modern in style; the animals more like those of today, but there were no more quartz filled cavities.

And now inasmuch as the aim of this geological narrative is to give, above and beyond the physical changes, the story of the life that the providence of God developed in this land of ours, it is needful that we seek to trace the beginnings of those forms that for so long a period were indigenous here in Oregon.

THE DALLES GROUP

One standing on the streets of The Dalles and looking southward will hardly fail to notice a well defined ledge of grey stone set against the hills a mile or so from the town, and extending westward three or four miles. It is a remnant of an old lake bed that once extended across the valley till its further margin set against the Klickitat mountain. What remains of this lake bed is today an unbroken level, although surrounded by many of the grandest exhibitions of volcanic and earthquake power, proving that no great violence has troubled the region since the waters of a quiet lake deposited its sediment there. A few years ago this rock was extensively used for building purposes at The Dalles, and in one of the building blocks, taken from the quarry, was found a well defined fragment of a metacarpal bone of a camel. This fossil became of added interest when a fragment of a very small radius came into the hands of the writer. The radius was found, among other bones, two or three miles from The Dalles by men searching for Gypsum in an eastern extension of the same quarry stone. The camel to which it belonged was a small member of the family, and represents an animal perhaps twenty-five to thirty inches in height.

There is a curious piece of geological history brought to the front in endeavoring to explain the circumstances under which this grey stone of The Dalles Group must have been deposited in those far away times. It, today, represents the bottom of a former lake. It is two hundred and fifty or three hundred feet above the present level of the Columbia River. The river has, in excavating its present bed, washed away the whole of that lake bed excepting this stone remnant, and in addition has worn its way through over two hundred feet of solid basalt, in reaching its present level. The east and south borders of this lake sediment are concealed by a covering of glacial deposit, under which it may be traced eastward two or three miles. It was from this eastward extension of the rock that the small fragment of radius was found, and from this same grey stone Mr. D. H. Roberts obtained the distal end of a well defined metacarpal bone of a larger member of the camel family.

THE DALLES GROUP

From this same locality the writer, many years ago, made a collection of fossil plants which passed into the hands of Dr. Newberry. In this collection was a specimen of birch; and a beautiful branch of acacia, the leaflets all finely outlined upon the grey stone and the branch carrying three or four large thorns so distinctly impressed on the rock as to give a vivid impression of its place in plant life. Besides these, there was an intensely interesting group of oak leaves indicating a range of four or five different species.*

*Note. The age of The Dalles Group is not yet positively determined, although Knowlton, the Paleobotanist, thinks the fossil leaves indicate the Upper Eocene age

See Fossil Flora of the John Day, Bulletin No. 204, U. S. Geological Survey.

CHAPTER VII.

LIFE OF THE LOWER LAKE REGION

OREODONS

The Oreodons, a large and very interesting group of mammals, now entirely extinct, were very abundant in the days of the lower lake of the John Day Valley. They had the molar teeth of the modern deer, the pre-molars or side teeth of the hog, and the incisors of the carnivors. They ranged in size from the stature of the coyote to that of an elk. The type of this animal is finely represented by the head figured on Plate VIII. The characters that mark all the fossils of the lower lake sediments are well shown in the teeth and bones of this head, which are dense and heavy, the teeth finely preserved and glistening with the luster of agates. Aside from the relation of this head to the life forms of the Miocene period, it is in itself an object of real beauty. And yet it is not as a piece of sculpture that this fine stone head exhibits its true rank. It is the added story one reads in its geological environment, for it is primarily a piece of the Blue Mountains of Eastern Oregon, and as such its rock material represents the fine sediment of a deep lake into which it was washed by a flooded mountain stream. Not a tooth of the forty-

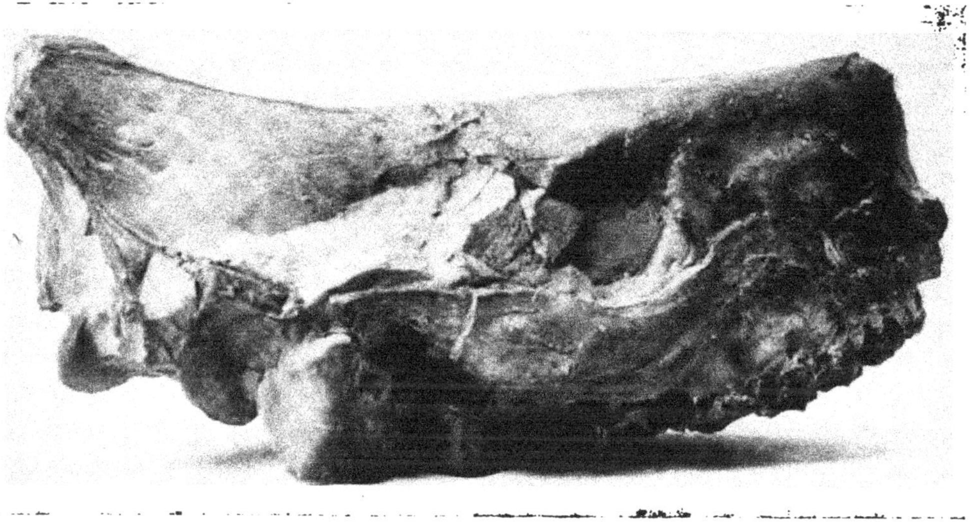

Rhinoceros Head, Side View. *Plate XIII.*

four this head contains, but has a measurable nerve cavity filled with some form of quartz which could only penetrate that cavity in a liquid state. And this liquid condition of quartz could only have occurred under great heat and pressure, two of the agencies by which the whole mass was changed into rock. Think these thoughts, then look again into that mute face.

The head figured on Plate IX. is in some of its features in broad contrast with the one just described. It belonged to an animal about the same size, but of broader, almost triangular crown, yet in the number and position of its teeth completely conforming to the Oreodon type.

The head figured on Plate X. has a feature that points to variation in another direction. The nasal bones are longer and take a more elevated direction. There is, too, a change in the lengthened tips as if these were still farther extended by cartilagineous endings, a provision that might well remind one of the upper lip of the tapir. The ruminant molars of the Oreodon would indicate that it often fed on grass and reeds, but the great spread of the canine teeth of some species would make it very difficult to use its incisors in nipping the grass. Just here an extended upper lip might be used to draw the grass inside the line of its canine teeth.

Another handsome Oreodon head in the collection of the writer has some new features. The canine teeth, though well developed, are drawn within the really graceful outline of the cheek, giving full play to the incisors.

On Plate XII. is figured the lower portion of the head of an Oreodon no larger than that of a fox, marking, perhaps, the extreme range of the family.

In looking over the cases that contain the fossils of these lower lake beds with the thought of the comparative numbers of these large mammals in mind, the figures were found to be: rhinoceros five, horses seven, dogs four, cats three, peccaries and hogs five, Oreodons twenty; and these numbers fairly represent their relative frequency in the fossil beds. It will be seen, then, that in the frequency of their occurrence as fossils, one sees the importance of their place in the life record of the Miocene. Paleontologists ascribe to their skeletons some features of the deer, others of the camel, and with great unanimity set him down as a ruminating hog, and his anatomy as that of a comprehensive type.

One can scarcely study such a form as he loosens fragment after fragment from a crumbling hillside, without a conviction that the law of lineal descent, with the holding power of heredity, and the directing power of an all comprehensive plan, entered together into its creation. The Almighty's work of creation as recorded among these Shoshone hills of Miocene times was through the agencies of mountain streams, of uplifting forces, of scattering seeds, of the nurture of plants and animals, and of gathering into this favored region the life that this same providence had prepared for this western Eden of the Miocene.

THE RHINOCEROS

That in the sediment of a large lake with a semitropical climate, the fossil remains of the rhinoceros should be found abundant, is not surprising; but even this consideration hardly prepares one for the frequency of their occurrence. This relative abundance would be explained by the fact that the life and death of the rhinoceros are both in lake bed environment. The teeth are large and showy, generally finely preserved and attractive as fossils.

On Plates XIII. and XIV. the reader will find a good head with teeth in place and in fine condition. This head measures seventeen inches in length. The lower molars of the rhinoceros may always be known by the shape of their crowns which takes on the likeness of the letter L, as may be seen on Plate XV. The clumsiness of these bones and teeth has given better opportunity for infiltrated quartz, and the agate luster that follows covers the whole surface of the fossil teeth.

As compared with the fossils of the horse, one is surprised at the small amount of variation of these bones and teeth, a fact that has an explanation in a comparison of their contrasted environment, the horse being essentially an upland mammal surrounded with wide varying of crag and mountain, while this type of rhinoceros, like the rhinoceros of the present time, was shut into the dull uniformity of the swamp and lake shore.

It is now known that the tapir, the rhinoceros and the horse had a common ancestry, living perhaps in the early Eocene. And we find the very interesting discovery made by trained paleontologists, that "The rhinoceros tribe had deployed into three notable branches, * * * a true lowland form ancestral in type to the existing family, another aquatic, a third, an upland, horselike, running form, Hyracodon." But from some cause the "running and swimming branches had dropped away," and only the lowland, usually hornless branch of the family has been found in the Miocene.

ELOTHERIUM

The Suidae, or hog family, is represented in these lower lake beds by several genera with different measures of the features of that family. One of these, represented on Plate XVI. by a jaw and several molars, under the name Elotherium or Entelodon, is perhaps the largest of the family. This old-time companion of the rhinoceros reminds one in many ways of the hippopotamus, to which it was closely related. Its promient canines, unlike the angular tushes of the hog, were well rounded,

Rhinoceros Head, Upper Molars. — *Plate XIV.*

and are often found in the rocks of this Miocene lake. Many years ago, Dr. Leidy described this Entelodon from specimens discovered in the Bad Lands of Nebraska. There, too, they were found in lake sediment in company with the rhinoceros and oreodon.

BOTHRIOLABIS

Another member of the hog type was represented in Oregon by the Bothriolabis. There were two separate species which fairly represent the peccary of today. The larger of these, given on Plate XVII. looks quite hog-like; the other one has a symmetrical, handsome head of a smaller Bothriolabis.

Besides these finely preserved skulls and teeth, there have been gathered many fragments of the same type, showing peccaries were abundant in those lower beds of the Shoshone.

SMALL RODENTS

Rabbits, squirrels, rats and other small rodents in minute fragments, and sometimes in well preserved specimens, are found in these lower beds.

LEPTOMERYX

In the same rocks from which we gathered the teeth of these rabbits, we found some teeth that would be at once pronounced as a miniature form of deer, but about the size of the rabbit. There is such a small animal, the Leptomeryx, known as the Musk Deer in southeastern Asia; but the knowledge of that fact hardly prepares one for the sight of its fossil remains in the sediments of our lower lake, where they are often found.

Dr. Leidy also described the fossils of this beautiful Musk Deer from the Bad Lands of Dakota.

FELIDAE (CATS)

On Plate XX. is found a fine head of a primitive cat that represents an animal about the size of an Oregon Cougar, only the skull is a little narrower and the teeth longer and more slender. This head was found on the North Fork of the John Day River, and represents an animal that was common in the days of this old lake.

The fierce saber tooth cats, which were very abundant in this region, had long sword-like canine teeth serrated or saw-toothed on one or both edges, making a most formidable weapon for warfare. The original of the engraving mentioned above was closely allied to these, and was a fierce and destructive animal. It is believed these saber tooth cats were important factors in the evolution of the horse from the slow little animal that plodded quietly on twelve small hoofs to the fleet footed racing type of later times.

CANIDAE (DOGS)

The dog type of this Miocene period was represented by many species, one of which is figured on Plate XVIII. by the fragment of an upper jaw with its three most characteristic molars. They are well preserved and so distinctly dog as to leave no doubt as to their identity. These teeth would indicate an animal about the size of a Newfoundland dog. The name Temnocyon meaning cutting dog has been given to the genus; this with his specific name, ferox, would indicate a ferocious animal. The fossil was found along the North Fork of the John Day River by a man who was in the writer's employ in 1894.*

The small dog's head given on Plate XIX. was taken from the fossil beds of Bridge Creek, several years ago, by J. W. Cusick, an Albany banker. The writer once found in these lower beds the molar teeth of a fossil dog, which were twice the size of our largest Newfoundland, but the term dog, as here used, must be understood as covering a wide range of carnivorous animals, such as wolves, jackals and foxes. It is said there were at this time more than a dozen species of wolves living in our Shoshone region; in fact Eastern Oregon was very rich in animals belonging to both the dog and cat families. Many of them were described years

*Note. This was the fossil dog to which the author originally gave the name Canis Shoshonensis. But it has been found that the name Temnocyon ferox had been previously given to another fossil representing the same genus and species.

ago by Professor Cope, Dr. Wortman and others of the United States Geological Survey. Dr. Merriam of Berkeley, California, has reviewed these older records and after adding new material, has revised their classification.*

MESOHIPPUS

The mammalian division of animal life seems to have been started in the world with the five toes growing from a wrist or carpus of the forelimb. So plainly acceptable is this statement that even the whale has this carpus of the mammal hidden in his fore flipper with its complete attachment to rudimentary toes. Each toe is held to the wrist attachment in mammals by a long slender metacarpal bone, originally five to each hand; and the divergence of mammalian families was plainly secured by divergent treatment of these metacarpal bones. The whole five were made to start off with nearly equal development for the bear, the cat and the dog. These metacarpals were started with the central one dominant and the rest subordinate for the horse family. Two were evidently started together, with nearly equal development, for the needs of the ruminants, the camel, the ox, the deer, sheep and goats. Later yet, the completion of the plan calls for a mammal with one metacarpal to serve the purpose of a single digit, and to secure

*Note. Carnivora from the Tertiary Formations of the John Day Region by J. C. Merriam.

this, natural causes were set to work to make dominant the central metacarpal and slowly cast aside the subordinate ones.

Scarcely any lesson in Paleontology has in it more of interest than that of the intermediate stages of progress that connect the functional hoof of the Tertiary horse with the rudimentary splint of the living horse of the present time. Here in the Shoshone land of our story—the Eastern Oregon of today—are the archives of a large part of this horse history of the past. In short, in these Shoshone rocks we see glimpses of God's creative work of the ages in transforming a five-toed animal to one of a single digit. It is the revelation of this creative process that makes the fossil horse of Oregon so full of scientific interest.

On Plate XXI. is a good illustration of the metacarpals of the horse in one of the stages of this transition. The condition of these metacarpals in the living horse, is seen on Plate XXI. (1) The lower ends cast loose, only the upper ends articulated, while in figure (2) these lower ends are articulated to phalanges and these in turn to hoof cores. The enveloping sediment that has preserved these bones in place for untold ages is now changed to rock, showing the exact position of the distal end of these rudimentary hoofs.

The form of horse that most abounded here in the early Miocene period, was the Mesohippus. It was a genus of three or four species, varying in size from that of a Newfoundland dog, twenty-five to twenty-seven inches in height, to that of a small donkey. In spite of the many close resemblances between the skeleton of this Mesohippus and that of our living horse, there were a few features in which they differed widely. In the molar teeth of the living horse the divisions of the crown pass into prisms nearly the whole length of the tooth, while the corresponding molars of the Mesohippus are planted in the jaw by fangs or roots. Another feature in which they differ is in the divisions of the foot. In Mesohippus there were three continuous sets of bones in each lower leg joined to as many separate hoofs; while in the living horse, two of the hoof attachments are only rudimentary, their function lost. The result is that the living horse has but one hoof, while our Mesohippus had three functional hoofs for each foot.

The teeth of Mesohippus are wonderfully preserved, not in outline only, but so completely silicified as to carry the luster of agates. Many of these fossils indicate a really beautiful little animal of graceful outline, about the size of an antelope, bringing to that early period a truthful prophecy of the highest type of our modern horse. And so abundant were they on the hills of Shoshone that fragments of skeletons are found in nearly all its fossil beds.

On Plate XXII, is a good figure of the lower front teeth of this animal from the lower beds of the John Day. It was of this handsome fossil that an experienced stableman once exclaimed: "Full mouth, five years old—past. Horse? By George! It is."

On Plate XXIII, is a good figure of the molar teeth of the Mesohippus from the same region.

In the continued wear and tear of the cliffs of these fossil beds under frosts and storms, the fragments that survive erosion longest, and therefore accumulate on the drainage surface, will always be the teeth and the thickly enameled articulating surface of the large bones of the skeleton. The ends of the femur, the ends of the tibia, and especially the ends of the radius are so distinctly marked in the horse skeleton that they quickly catch the eye of the collector.

While the horses of the early Miocene period of the Shoshone region differed in size, it is noticable that the different parts of their skeletons retain their relationship in size and general form, so that any bone of any species found today would at once be seen to conform to the type and show the size of the animal to which it must have belonged. The mistaking of the bones of the young of a larger species for the adult of a smaller one need never occur, for the bones of the young carry with them the marks of their tender age.

LIFE OF THE LOWER LAKE REGION

Owing to the explorations conducted by the American Museum of Natural History and others in the fossil fields of the United States, many valuable fossil horses have been discovered, including several almost entire skeletons and a great wealth of less perfect specimens. This accumulation of new material made it necessary to thoroughly revise the old classification and to change some of the older names. Some of the fossils formerly called Anchitherium have been transferred to the genus Mesohippus, which was an earlier type of three-toed horse, still smaller than Anchitherium. In addition to its three front toes, it carried a splint, the useless remnant of a fourth toe. The two small specimens given on Plate XXII, showing the lower front teeth of the little horse, have been pronounced by an expert Paleontologist Mesohippus and not Anchitherium.

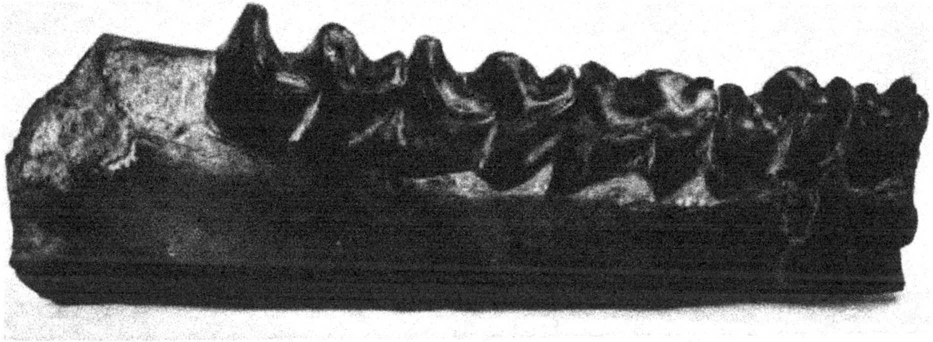

Rhinoceros Lower Jaw. *Plate XV.*

CHAPTER VIII.

LIFE OF THE UPPER LAKE REGION

TRANSITION

Before dismissing the history of this lower Miocene lake, let us ponder a moment on the evidence we have of its long duration. We have a sediment of mud three thousand feet in vertical thickness, by far the greater portion of very fine materials, indicating a slow rate of deposit; other portions coarser, indicating a more rapid deposit; but an average accumulation of not more than a small fraction of an inch per year, and this built up to the thickness of three thousand feet. Yes, this work covered an immense period of time.

Now it was this vast accumulation of lake sediment changed to rock, and every yard of it teeming with fragments of organic forms that tells the story of its times. But more. This great lake bed was slowly and unevenly elevated, till in some places it was tilted to an angle of several degrees from its original level. It was slowly emptied of its covering of water till an extended valley, only, remained to mark its outline—but the early Miocene lake had gone forever. The tiltings of the bed of the old lake, its foldings, and crumpled condition, give striking evidence of the great force that mark its passing away. Another fact of like import calls for a word of notice—the occurrence at irregular intervals of dykes of trap showing masses of lava that were forced up from below, filling great orifices in the hardened mud of the lake bed and speaking of nothing if not of force and heat. It was in this cloud of confused and struggling forces that the early Miocene record closes.

A large part of what is now Eastern Oregon then began a new series of changes on the surface of the upturned ruins of hill and vale. And upon this uptilted mass of confusion, rain and snow, heat and frost, a new structural surface began, and continued its reshapings until a drainage system was worked out for another geological period. How long this period of reconstruction lasted is not recorded; its very stages of progress have long since been blotted out. For aught the record would indicate it may have lasted as long as the whole time of the lower lake history; and throughout this reconstruction period, the hills and plains of Shoshone may have richly abounded in mammalian life, only lacking nature's favorite mode of recording the life of the hills, a good deep lake bed into which mountain streams might in time wash their life remains.

The new lake period at length opens as a result of these reconstruction changes. The old Shoshone lake had covered a large part of what is now Eastern Oregon; the new lake system consisted of a number of more restricted bodies of water. One of these, the earliest published, may be found along the upper John Day Valley, beginning where Cottonwood Creek joins the John Day River, and extending several miles southeastward, the channel of the river being excavated through it. It will be remembered that we called the lower Miocene lake beds the lower lake deposits. These of the second group we will call the upper lake deposits. This upper John Day lake sediment, then, is to be with us the type of the group.

HIPPARION

If we look for the record of the horse type in this new chapter, we shall find a wonderful increase in the variety of its forms. The one that attracts our attention most from the frequency of its occurrence and beauty of its fossils, is that of the Hipparion type. And as it occupies a midway position between the Mesohippus and living horse, if we compare the Mesohippus with the Hipparion, we shall have the amount of change through which the horse passed from the Lower Miocene to the Late Miocene or early Pliocene age. And carrying the work farther, if we compare the Hipparion type with the present horse we shall have the amount of change through which the horse has passed from the Hipparion stage to the present. We find the simpler teeth of Mesohippus have in the Hipparion type taken on a more complex form in the prisms of their enamel, see Plate XXV., and that the molar teeth of the Mesohippus were, through a large part of their trunk, divided into spreading roots, the body part of the tooth remaining short, while in Hipparion, the molar teeth have small short roots and a large increase in the depth of the body of the tooth—so much so as to remind one at once of the teeth of the living horse.

That we may extend our comparison to the feet of these Tertiary horses, let us study the fossil shown on Plate XXIV. Here is a foot of the Hipparion type, consisting of three metacarpal bones with corresponding phalanges, the central one much the longest and by far the most symmetrical; all three are articulated both above and below, above to the carpus or wrist, below to the phalanges or toe bones. It has been stated in the last chapter that the Mesohippus had three useful hoofs for each foot, but this Hipparion fossil shows the two outer metacarpals so shortened that their hoofs were plainly carried out of reach of the ground as he traveled, giving evidence that the horse of this later age was losing his second and fourth toes. If we now compare the same fossil with the corresponding bones of the living horse on Plate XXI. we shall find a very marked relative increase of the central metacarpal and a great relative diminution in the outer bones until in the modern horse we see the total retirement of the second

Entelodon Head.

Plate XVI.

and fourth metacarpals except as useless splints. This change occurred between the time of the Hipparion type and that of the living horse. Or these facts stated a little differently—formerly the metacarpal bones held the toes to their work, and for this purpose were articulated below to the phalanges or toes; now these toes are cast off and the metacarpals are only articulated above. Formerly the middle or third metacarpal differed but little from the outer ones in size, strength and symmetry of form; now it alone can be considered, for all others are rejected from their former functions except this third or center bone. The rest are but the useless tools of former working forces.

Professor Osborn, of Columbia University, has called the American species of Hipparion by the new name of Neohipparion and writes of it as follows: "Neohipparion was proportioned like the Virginia deer—delicate and extremely fleet footed, surpassing the most highly bred modern race horse in its speed mechanism, and with a frame fashioned to outstrip any type of modern hunting horse, if not of thoroughbred."

PROTOHIPPUS.

Another horse whose remains are abundant in the upper lake bed is designated by Paleontologists as Protohippus. He resembled Hipparion in size and in the length of the small side toes, for in both animals the second and fourth toes had become useless remnants of an older type. Protohippus is believed to be in the direct line of descent of the modern horse, while the Hipparion type soon died out, leaving no descendants.

The Protohippus was also found in Eastern Washington. A banking house in Ellensburg was quarrying stone from a neighboring hill for their new building, when a workman found some teeth imbedded in solid rock. These were sent to the writer by W. R. Abrams, asking if they had not found a fossil horse, and they proved to be teeth of the late Miocene horse Protohippus.

CAMEL.

Next to the horse in the variety of interest to which the subject appeals, stands the camel type among mammals.

A small primitive member of this family, the Poebrotherium, lived in the Shoshone region on the borders of the lower lake with the Oreodons, the Mesohippus and Leptomeryx, but camels were more common and, many of them, of larger size during this later period of the upper lake. The ulna and radius of one of these may be found on Plate XXVI. where it is compared with the corresponding bones of a large recent dray horse, so that the reader may judge of the range in size. This animal must have been fully as large as the present Arabian camel. Another, apparently about the size of a goat, and perhaps akin to the South

American llama, is represented by a portion of the head and teeth of a very good fossil from the upper John Day Valley. Three phalanges of the camel were also found in digging wells near Walla Walla.

The camel was not only well represented at this time in the Pacific Northwest, but probably roamed in great numbers over the whole United States, for it is now known that the camel was a native of North America. It is believed that a little later than the period of our upper lake, some of the family found their way over the newly elevated land bridge into South America, and became the ancestors of the llamas and vicunas of that country, and another migratory band crossed overland near Bering Straits into Northern Asia, or perhaps by way of Greenland and the British Isles into Europe. Thus the Old World was supplied with the camel, which, in historic times has been such an important factor in the commerce of the East.

CANIS (DOG)

The fine head of a dog, which the reader will find on Plate XVIII., was badly broken when taken from the rocks, but with a great deal of labor, it was reconstructed. It is one of the few fossils the writer has ever attempted to name, but when the restoration was complete, it looked so dog-like that it seemed unwise to send it away from home to be named—a procedure that in other years had cost the writer many attractive fossils, so it was decided to find for it a name at home. And inasmuch as this dog could not go to the large cities of the world where the record of new species was kept, it seemed fitting to treat him like a country cousin and call his name Canis rurestris.*

*Note. After making a careful study of this fossil, Dr. Merriam, of Berkeley, California, has found that it differs somewhat from the genus canis and has substituted the name Tephrocyon rurestris for Canis rurestris.—E. C. M.

Bothriolabis Head. Plate XVI

CHAPTER IX.

SEALS.

While the ocean still filled the Willamette Valley, and its tides ebbed and flowed between the islands of the Coast Range, there lived along the Oregon Coast a primitive seal.* His fossil head was discovered some years ago embedded in sandstone, and is said to represent the most ancient type of seal yet known. It is a rare and valuable specimen, and throws much light upon some of the unsolved problems connected with the origin and development of the seal family. Some scientists have thought that modern seals are descended from primitive bears, but our Miocene fossil is almost as old as the beginning of the bear family, and yet the seal and the bear differed then quite as much as the modern bear differs from the seal of today.

Others have questioned whether our common seal (the Phoca) did not have a different origin from the other members of its family. But in our Oregon fossil we find many of the characteristic differences between the sea lion, the fur seal and the common seal all blended in one ancient animal, thus strongly indicating a common ancestry for the whole group.

*Note. See A New Fossil Pinniped, University of Oregon Bulletin, May, 1906.

We feel no hesitation in assigning this fossil to the Miocene for, owing to change of level on our coast, Marine Pliocene deposits are very rare in Oregon. And in the vicinity of the fossil bed in question, there are but two formations: first, the elevated beach of the Pleistocene which, with its capping of recent sand dunes, stands nearly one hundred feet above the level of the present beach; and second, there is a solid base of Miocene sandstone dipping gently toward the sea. It was in this sandstone that the seal head was found, and the rock is rich in such characteristic Miocene shells as Arca, Nucula and Glycimeris.

That the sea lion, fur seal, walrus and common seal or phoca are closely related to the dog, bear, tiger and other flesh eating land animals, seems to be a fact too well established to be questioned. But our Miocene fossil teaches us that the seal family was probably not derived from any of these modern land animals, but that we must look for the first seals before there were any true wolves or bears or tigers, so far back in time that all of these forms were still blended in the more ancient and generalized type of mammals. This is the theory long held by Doctor Wortman, the paleontologist, and he believes our Oregon fossil strongly confirms his theory. And yet, while the study of our fossil seal is intensely interesting, the problem of its ancestry still remains unsolved, and one of the strongest convictions left upon the mind is the antiquity of its type. For the structure of our seal of the distant Miocene is perhaps as far removed from his probable land loving ancestors as from his sea-faring relatives of today. And who can stand at this Miocene way station in the evolution of seals, and look first backward toward the ancestral stock, then forward down the intervening ages to the present forms, without feeling the grandeur of the slow, persistent, unceasing progress of creation?

CHAPTER X.

THE PLEISTOCENE

THE SILVER LAKE GROUP

After the upper Miocene lakes with their rich variety of animal life and their grand forests had passed away, we find a still later body of quiet water filling the upper John Day Valley, only a narrow Pliocene lake—the last of the wonderful series that, with many interruptions, had persisted in this region through long geological ages. For the period of lake history was giving place to the great river system of the Columbia and its tributaries.

The Pliocene age was a time of increasing altitude and of mountain building. The broad dome of the Cascade Range had begun erecting its superstructure. For the lofty peaks, the rugged crags, the deep canyons, the grand volcanic cones were largely the work of Pliocene and early Pleistocene times. Because the sources of the rivers were so greatly elevated, they became swift mountain torrents, carrying with them increased eroding power for cutting down the canyons and mountain gorges through which they flowed. We know that throughout the Pliocene, herds of native horses and camels roamed over the hills and valleys of our

Northwest Coast, but the many quiet stretches of lake and river that had covered and preserved the animal life of earlier times seemed wanting, and well defined Pliocene fossils are rare in Oregon. But if we turn to the next period, we find a rich chapter of life in the "Equus Beds" of the Silver Lake region, which was formerly assigned to the Pliocene, but is now classed as early Pleistocene or preglacial time.

In 1876, Governor Whiteaker, while camping in central Oregon, in the neighborhood of Silver Lake, noticed some fossil bones on the surface of the open prairie, and shortly after this brought some fragments to the writer of these pages for examination. The Governor was soon convinced that he had discovered an important fossil bed, and the next summer, by kindly furnishing a team and sending his son as guide, he gave the writer the pleasure of visiting this Silver Lake country.

Considering the narrow area of this fossil bed, a surprisingly large number and variety of fossils were found and so brought to the light of scientific report. The last part of the journey took us through a monotonous dead level covered with sage brush, until finally we reached the home of a ranchman on the shore of one of those strange alkali lakes whose flats are at this season covered with a thick inflorescence of alkali. Here we left our wagon and the next morning started on horseback for the fossil beds. After traveling about eight miles, we saw, from the eminence of a sand dune, an apparently circular depression four or five miles across, in the lowest portion of which was a small pond, or lake, surrounded by grass and *tule* rushes. Perhaps two miles to the leeward this depression was bordered by a line of sand dunes, unquestionably formed from sands blown from the bed of the lake that once occupied the whole of this depression. It is the blowing out of this sediment which exposes the fossils buried in the depths of the old lake. Here we staked our horses and went to work. We found many fragments of elephant bones, a fine collection of bird bones, the bones of a large horse, a large camel and the remains of a smaller animal of the camel family, the Auchenia shown on Plate XXVII. which Prof. Cope named in honor of Governor Whiteaker, Auchenia Vitakeri.

Judging from the uniformity of its surroundings, one is found unavoidably thinking of an extensive lake sediment, of which this fossil lake is only a very small portion. The original lake probably included Silver Lake and Klamath marsh with its surroundings, and perhaps Summer Lake and an extension Eastward over the present Harney and Malheur Lake regions. These waters were lowered to their present level by evaporation in excess of inflow. The mineral left behind accumulated in the process until it covered the face of the pond like snow. These waters must have varied in extent at different periods. From one spot, the writer could mark an extent of not less than sixty miles from east to west and fifteen to twenty from north to south, with a variation of surface scarcely reaching what an ordinary eye would call thirty feet. And this whole extent was water-covered during the life of the elephant, as is proved by his remains. The portions of this extensive lake bed, which remained latest, caught most of the animal bones buried in its mud. This special one we visited continued sandy and when dry, its contents were laid bare by drifting winds.

The fossils of these Silver Lake beds were found often lying on the surface, bare of any covering. The sands and dust that had covered them were blown to the leeward where they now lay in extended dunes; and this uncovering and drifting process was still visibly going on. Among these fossils we found many arrow heads of obsidian, such as were used by recent Indians. We found, too, lying among them, many fresh-water shells of species now living in the waters of Klamath Marsh. Shells and arrow heads were, like the fossil bones, entirely uncovered lying upon the surface of the ground. If the sands, the fossils, the arrow points and the fresh water shells were all of the same period, and the fossil bones were early Pleistocene, then the arrow points were fashioned before the glacial age and men inhabited the surrounding hills in the early Pleistocene period. But the mixture of these facts may be due entirely to the simple law of gravitation, for both the arrow points and the recent shells may have settled down among the fossils as the dust and sand upon which they rested were gradually blown away.

The rarest and most valuable fossils collected on this expedition were the eighty bird bones which were unusually perfect and in a fine state of preservation. Later they were sent to an Eastern scientist for classification and never found their way home. Fortunately they had been examined by an expert in the study of fossil birds and to him we are indebted for much interesting knowledge of the life of the region.* Dr. Shufeldt tells us that here at Fossil Lake, lived five species of gulls, two of terns, eleven species of ducks, four of geese, one of which "must have been nearly as large again as our common wild Canada goose." There was also a large species of swan named for Governor Whiteaker, Vitikeri. There were great horned owls, blackbirds, coots, herons, crows, eagles, grouse, prairie hens and a great cormorant. "But the strangest figure upon the scene among the birds was a true flamingo."

And found with these bird bones, Dr. Shufeldt also enumerates the fossil mammals as including a great sloth as large as a grizzly bear called the Mylodon, and at least four kinds of camel, ranging

*Note. See "A Study of the Fossil Avifauna of the Equus Beds of the Oregon Desert," by R. W. Shufeldt.

Canis rurestris. Plate XVIII.

in size from a modern camel to the smaller llama. There were also bears, coyotes, rabbits, gophers, otters, beavers, a mammoth elephant and three species of the modern type of horse or Equus. These "Equus Beds" are considered among the richest yet discovered of the early Pleistocene period. It is interesting to note the variety and great abundance of the fossil remains of the camel and the horse whose families are represented among the oldest fossil mammals of the Northwest. And here we find both lines persisting with unabated vigor even to the beginning of the glacial ice.

THE TOUCHET HORSE.

While the finest fossils of the modern type of horse have been taken from the Equus Beds in the Silver Lake region, a still more interesting discovery was the finding of the Touchet Horse ten years earlier. In the spring of 1866, a new mining interest in Eastern Oregon, Washington and Idaho resulted in an extensive demand for miners' supplies along the line of Pend d'Oreille Lake. The merchants of Walla Walla, in an effort to secure resulting trade, opened a road from Walla Walla to Palouse Landing on Snake River. The distance along this road from the crossing of the Touchet to Palouse, was thirty miles without water. To remedy this need, the road company dug for water fifteen miles beyond the crossing of the Touchet. During the same spring, and in the interest of the same trade, Mr. Moody, of The Dalles, afterward Governor Moody of Oregon, was building a small steamer on Pend d'Oreille Lake, and on his way home encountered these well diggers of the Touchet wagon road. They had dug through gravel to the depth of eighty-six feet without striking water, but Mr. Moody found them examining some fossil bones they had just found at this great depth. They, thinking they were human remains, turned to him for an explanation of the mystery and he promised to carry some of the bones to The Dalles, where he told them he had a friend who studied such things. One of these fragments was found by the writer to be a remarkably well preserved specimen, shown on Plate XXIII. of a fragment of a radius of a horse quite as large as the corresponding bone of a good sized dray horse of today. Careful inquiry brought out the information that the region from the Touchet to the Palouse was nearly level, and the whole eighty-six feet of digging was through river wash. Here, therefore, was proof that when this horse lived, a lake thirty miles across and eighty-six feet deep stretched from the Touchet to the Snake, a depression that was slowly filled up to its present level by the river flow of the region. The same winter, the writer published these facts by a lecture in Portland, and the discovery was published in the Portland Oregonian. This Touchet horse, found in 1866, was so far as is known to the writer, the first fossil horse discovered in the Pacific Northwest.*

DURING THE GLACIAL AGE

During the Glacial age, when for long periods most of our northern states were covered by ice and snow, Oregon was not under a continuous mantle of ice, but had many independent glaciers of its own. Remnants of these still remain in places and the previous existence of others is proven by ice scratches, moraines and other evidence of glacial action found in many of our mountain valleys.

The present glaciers of Rainier, Adams, Hood, Jefferson, The Three Sisters, and Mount Mazama are only small remnants of greater ice streams that with steady grinding power plowed their way from these lofty snow peaks during the Glacial age. Lindgren tells us the Eagle Creek, the Elk Horn and Green Horn Mountains of Northeastern Oregon all had their glaciers. Russell writes of old glaciers in the Stein Mountains. A glacier was plowing its way over the hills east of The Dalles. There were glaciers in the McKenzie and Mohawk Valleys, the Willamette Valley had glaciers reaching to its borders.

Dr. Diller tells us that during the glacial period a grand snow peak towered above the present Crater Lake. This mountain, called Mount Mazama, once rivaling Shasta and Rainier in

*Note. The age of the Touchet horse is not well established. It may have lived during Pliocene times and so be earlier than the Silver Lake Group.

grandeur, was not only a snow peak, but an active volcano during the glacial age. Some of the lava that rolled down the sides of the mountain and cooled into volcanic rock was later scratched and scarred by the ice streams that crept slowly toward the valley.

When at last the long periods of cold had passed away and the more tender herbage, shrubs and trees had crept northward, the great lapse of time and the increasing struggle for existence had worked through natural laws to produce many changes in the life of the Northwest. The camel seems to have disappeared and the herds of wild horses to have returned no more. The forests had lost much in richness and variety of forms. Many genera that once flourished so luxuriantly have never returned to our Northwest Coast.

While the Pliocene and Glacial periods were times of increased elevation, with the passing away of the Glacial age there was a gradual sinking of the land. Not only did the Pacific States lose much of their western border recently acquired from the ocean, but the sea gained upon the land until the water stood several hundred feet higher upon the coast than it does today.

SURFACE BEDS

The field intended by the term surface beds includes all slight depressions of the surface producing ponds with sediment enough to preserve

Canis, a Smaller Type. *Plate XIX.*

bones and teeth washed into them, and also swamps and bogs into which large mammals often sank to their death, leaving their bones to such preserving agencies as might occur there. And inasmuch as the latest great surface leveling agency of the North Temperate Zone was that of the glacial ice, most of these surface depressions would date from glacial times and would, therefore, be properly designated as late Pleistocene. Furthermore, up to the glacial period, the horse and the camel were abundant here, and the question of their continuance in Oregon through glacial times is still in doubt, so that our group of surface sediments must provide the settling testimony on this question.

If the bogs, swamps and minor surface depressions furnish no horse or camel bones, then must it be accepted that the glacial cold or some other agency drove these mammals away or destroyed them. It is plain that the mammoth elephant got him a coat of fur and lived through the cold spell of the times. The fossils of this group of surface beds, such as the Mastodon, the Mammoth, the Broad-faced Ox and Mylodon, though deeply interesting in themselves, bring added historical attraction from the fact that a large part of their geological period overlaps that of pre-historic man.

THE SLOTHS

There were four of these great Ground Sloths in North America during the Pleistocene period, the

Megatherium, the Mylodon, the Megalonyx and the Glyptodon, some of these animals were larger than a rhinoceros, fifteen to eighteen feet in length and their teeth indicate that they fed upon plants and browsed upon trees. Their remains are rare in Oregon. The toe bones shown on Plate XXVIII. are from the banks of the Yamhill River, proving that one of these strange clumsy animals lived on the shores of the Willamette Sound in the late Pleistocene age. Remains of Ground Sloths have also been found in the Pleistocene beds of the John Day Valley, and in the Equus beds of the Silver Lake region.

It is believed that they migrated from South America in the Pliocene period, coming over a newly elevated isthmus into the southern and western parts of the United States, but they were more abundant in South America.*

BOS LATIFRONS

The precursor of the Buffalo in Oregon was this Broad-faced Ox. His horns were longer and

*Note. The reader will find an exceedingly interesting article by Dr. Merriam, of Berkeley University, published in Sunset Magazine for October, 1908. He writes of the fossils recently found in the asphalt deposits near Los Angeles, California. The remains of the extinct Ground Sloths are very common, embedded in the asphalt, and the knowledge of their forms and habits revealed by these specimens and others referred to by Dr. Merriam as found in a Patagonia Cave, will add greatly to our interest in these extinct monsters. They are now known to have been covered, over at least a part of their bodies, by long coarse hair, while on the underside of the skin are embedded "many small pebble-like bones" serving as a "partial armor." One of the interesting items noted in the article was the proof that man lived in Patagonia at the same time as the Great Sloths.

stouter, and his bony forehead was wider than that of the Buffalo, measuring nineteen inches across the line of the eyes. His brain cavity was very small, scarcely as large as that of a three months calf of today. His skull was not only very wide, but unusually thick, being two and a half inches thick in mid forehead. The front of this skull is figured on Plate XXIX. This one was found five or six miles east of The Dalles.

THE MASTODON

The Mastodon and the Mammoth are both abundant fossils in Oregon. In some features of their physiology they are similar, as in size, in having trunks possessing great delicacy of touch, and in the development of tusks of great power and of pure ivory, but they vary widely in external appearance and are most conveniently distinguished by their teeth. The crowns of the teeth of the Mastodon are covered with rows of conical protuberances as seen on Plate XXVIII; while the Mammoth tooth as given on Plate XXX., is composed of many flat, transverse plates cemented together. The Mammoth rarely retained more than four perfect teeth at one time—one on each side of each jaw; while the Mastodon had a larger number in constant use. But they are closely related and both belong to the great elephant type of mammals.

THE SILVER LAKE GROUP.

MAMMOTH.

In the Mammoth the student of animal life will always find the farthest reach of mammalian development. The Shoshone region had its full share of Mammoth life, for scarcely a digging of any importance in Eastern Oregon is without its list of Mammoth fossils. Whether one considers the size of the individual, the length of its life or the time allotted to its geological horizon, or the prodigous numbers to which it reached, this great elephant stands at the head of its column. In its presence, our living elephant is but a degenerate offspring, indicating the rapid passing away of the race. There are scarcely any fossil beds of this surface division but have their Mammoth teeth as trophies of their excavations. The Mammoth tooth is not simply a bulky mass, but an admirably constructed and complex piece of work. The unit in its structure is simple—a single cusp. The cusp is united with other cusps to form a flat plate. The cusps are cemented together by layers of dentine, each period of the elephant's life being accorded a specific number of plates so that the first molar, the milk tooth, is cut at three months old and shed when the calf is two years old; the second, of eight or nine plates, is shed at six years; the third, of eleven to thirteen plates, is shed at nine years; the fourth, of fifteen or sixteen plates, is shed at twenty-five years; the fifth, of seventeen to twenty

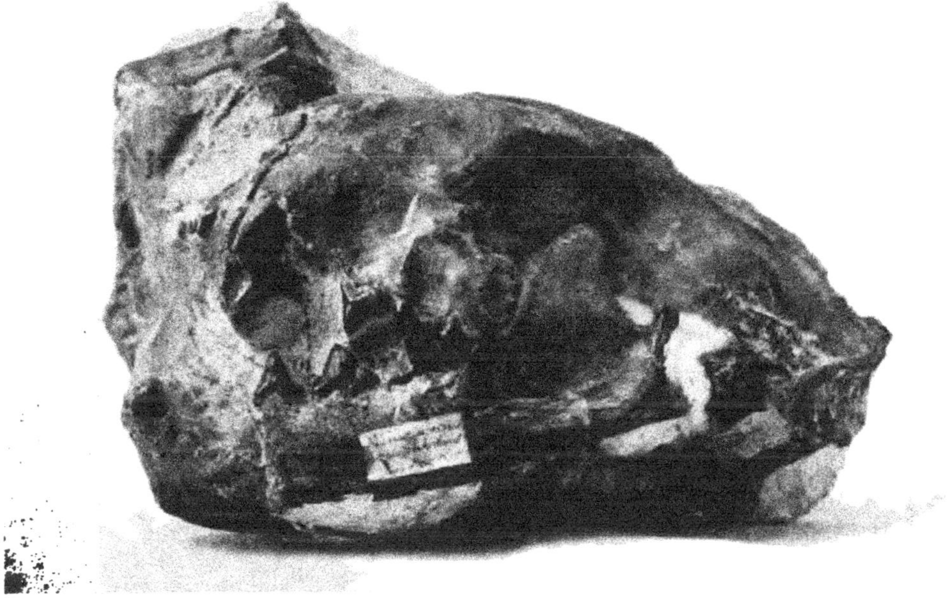

Felis. *Plate XX.*

plates, is shed at fifty; the sixth, of twenty-two to twenty-seven plates, is shed at a hundred. This sixth molar, given on Plate XXX., with its twenty-seven plates, measures nineteen inches in length and seven inches in breadth.

CHAPTER XI.

THE WILLAMETTE SOUND

PUBLISHED IN THE OVERLAND MONTHLY

NOVEMBER, 1871

The desire to study some of the evidences of the more recent changes of level along the coast of Oregon and Washington, with a view to compare and if possible to connect them with evidences of like changes in the interior, led to a visit to Shoalwater Bay, an inlet of the coast a few miles north of the Columbia River. June, on our Northern coast, is a pleasant season for such trips, and ours received its full measure of help from such accessories as bright sunshine and pure air above us, unmeasured wealth of form, color and fragrance below.

The ride from the cape at the mouth of the Columbia to Shoalwater Bay is one of the finest in the country. The road for the greater part is along the ocean beach, always strewn with the numerous wrecks of life cast upon its sands, and often presenting to the naturalist objects of rare interest. An abrupt turn of the road inland ends this finest of beach drives at ten or twelve miles from the cape. A short distance through woods of spruce and pine, thickly undergrown with a rich variety of flowering shrubs, and the road opens upon a fine view of Shoalwater Bay at the pleasant little town of Oysterville. The general outline of the bay is in sight from this point. The bluffs that define its shores appear, seen northward, fifteen or eighteen miles away, and in the direction of its southern extension, ten or twelve miles. At intervals along this whole shore line, one can plainly discern what in the distance appear as land-slides, but on nearer approach prove to be portions of the bluff shore undermined by the storm-surf, and in their present form showing fine sections of the strata of which they are composed. On examining these more closely one sees a bank, not of common earth but disposed in stratified layers of sediment, once evidently continuous over the whole region and of nearly uniform thickness, now worn away above into a rolling surface, yet showing everywhere a fine persistence in the old water-lines that ruled its formation. Buried in this mass of sediment, and occasionally cropping out in exposed sections, are vast beds of sea-shells. So completely do these represent the life now around them that when an apparently exceptional form does appear, memory at once recalls having seen it somewhere on the coast. And yet, identical in species as these shells unquestionably are with those now living in the surrounding waters, the two sets of conditions are separated by the whole import of the term "fossil." The waters that buried there those fossil shells, and covered them with one hundred or more vertical feet of ocean sediment, were waters that so defined our northern coast as to give it a far different outline from that of its present geography.

In some of these bluff exposures their past record is read in masses of buried forest trees; trunk, leaves and seed so buried in clay and so well preserved that the spruce cone, fragile at all times, is scarcely discernible from one of last year's fruitage drifting in the neighboring waters. From these vegetable remains, as from those of the shellfish, the same truths are taught; for the trees are the same in kind as those growing on the bluffs one hundred feet above them, while the waters that covered them there with one hundred feet of sediment have passed away. The fossil story, then, that may be read, is linked to our own times by the sameness of vegetable and animal life and separated from ours by the passing away of the agencies by which the records were written. It is useless to ask, How long ago? There is no chronological record legible here. Future discoveries may connect these things with human story. We may not attempt this now.

The lowest marine remains of these bluffs plainly prove that when they lived, the waters around them were at, or near, their present level. They were species that love shoal water, and they are "in place" where found. The oyster is very abundant among them, and the shells of most of them are neither broken apart nor water-worn as they would be if drifted here from some other locality. They evidently lie here as fossils on the same bed they occupied while living; and oysters then, as now, rarely bed in waters more than a few feet in depth. The common cockle—another lover of shoal water—is also abundant among these remains, and like the oyster, lies fossil where it lived, the opposite valves often occupying the very positions, relatively, that they held while living. So, too, with the members of the clam family; whether Mactra or Solen or Venus—all are evidently in their native beds where they lived and died. We conclude that when these shell-fish lived, the surrounding waters held nearly their present level.

Another truth plainly taught in these stratified bluffs is this: The waters here became afterward much higher, or speaking more exactly, the land became much lower. There must have been a change of more than one hundred feet, for a stratified sediment of one hundred feet in thickness as now seen in some of these bluffs—that for instance near the North River—would require more than that depth of water to place it there; and this sediment is so fine in material as to warrant the conviction that it once existed evenly distributed over the whole region, bay and all. The upper layers, too, have in them the finest materials and the fewest fossils; both facts indicating increasing depth of water as the upper beds were deposited.

Yet another plain truth is legibly written here: the changes indicated in depth of water over the place were quiet changes. Any sudden catastrophe would leave signs of violence and consequent strong current; but nothing of the kind appears here. The fragile cone of the spruce tree, buried in that sediment, is found today among those shells as little marred by time as the shells themselves. The line of deposit along the sheltered bay, just as it was at its deepest stage of water, is now as unbroken as it was then. Neither the violence of earthquake nor the suddenness of deluge has left any trace of such agency to disturb the conviction one feels that the changes indicated there were quiet ones, covering a long period of time, yet scarcely disturbing the quiet order of life over which they presided.

That every inlet on our northern coast has its group of facts of like import, there can be no doubt. Our line of thought needs only those that mark its extension to the Columbia River, and there the lessons gleaned from the bluffs of Shoalwater Bay reappear in all their clearness. A fine instance of this is seen in a bluff on the old Whealdon farm, just inside the cape. Several others may be seen along the streams that fall into Young's Bay, on the south shore of the river, and just back of Astoria. All these contain remains of animal and vegetable life linked to our shores and forests of today by identity of species and separated in our minds, from the present order of things, by the conviction that the agencies which placed them there have passed away.

In all this we are obviously studying only the lower limit of this latest of Oregon's geological changes. Where shall we look for its upper limit? In other words, how high did those waters rise above the present sea level? We might look for traces of its upper reaches in the remains of old sea-beaches in elevated places on the abrupt slopes of the hills along the coast; but in such exposures old beach-lines are but rarely preserved against the storms of a thousand winters, still less against those of tens of thousands. To find them and their records plainly legible, we must look to more sheltered localities inland. It will help us a good deal in our search for such shore-lines of the interior, to carry with us a theory that will point out the possible limits within which they may reasonably be sought. Will the facts we have gathered from Shoalwater Bay and the Lower Columbia, warrant us in forming such a theory? Let us see.

Stratified sediment of a hundred feet in vertical thickness—finer far in its upper layers than in those lower, and in its upper layers entirely devoid of marine remains while the lower ones are densely crowded with them—plainly indicate shoal water to begin the work, and deep water afterward over its highest layers. But the sediment itself is one hundred feet or more and deep water over its upper surface, equal to the requirements of its facts, could not be less than another hundred feet; thus making a total depth of at least two hundred feet above the present water level. Let, then, a depth of two hundred feet be our theory, and with this let us pass inland for facts to confirm it if true, to reject it, if false; and if confirmed, to trace by its help the outlines of that fine old Willamette Sound that may in the days of the Mammoth and the Broad-faced Ox have welcomed to its scores of sheltered harbors the ancient hunter who, in his canoe, if he had one, floated one hundred feet or more above the present altitude of the church spires of Portland and Salem.

But as we pass along, let us in imagination reconstruct the fine inland sea that two hundred feet of elevation in the waters of the Columbia must have made. We have first the noble entrance, like that of the Straits of Fuca, extending from the present site of Astoria to that of St. Helens, eighty miles or more in length, varying from five to twenty miles in width and over two hundred feet in depth. At St. Helens it spreads out into a broad inland sea, extending from the Scappoose Mountains to the elevated land east of the Willamette Valley. Like the Puget Sound of today, whose general outlines this old Willamette Sound strangely resembles, it was in its southern extension—over

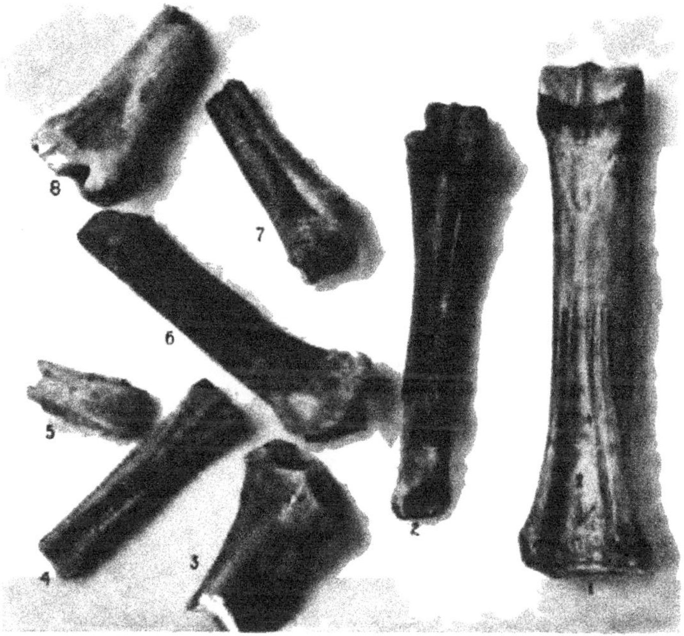

1 Donkey Bone.
2-8 Hipparion Bones.

Plate XXI.

the present valley, among elevated islands, deep channels, and land-locked bays—reaching from the Scappoose Mountains to Spencer's Butte, that it spread out its greatest wealth of scenic beauty. Our theory would make it cover the whole of the lower levels through which the Willamette now flows.

Let us trace this grand water system eastward, along the present course of the Columbia River.

We started, it will be remembered, from the capes with a theoretic elevation of the waters two hundred feet above their present level. The fall of the river, from the Lower Cascades to the ocean, may be stated at forty feet, the fall through the five miles of cascades, at thirty-five feet. Above this there are forty or fifty miles of narrow gorge through a mountain range, with slopes too steep for preserving old shore-lines and through which the river falls twenty feet more. Here we find the first open space east of the Cascade Mountains, in which the waters of that period—if two hundred feet higher at the capes than they now are—would have had an elevation above the present river level of one hundred and five feet. There was at this place a lake-like extension of the river seven or eight miles wide and fifteen to twenty long; and into this a semi-circular system of streams, six in number, brought a continued supply of sediment—sand, clay and gravel—and buried, year after year in its strata along the margin of that lake, the record of the passing events of the time. Now, manifestly, at whatever level we may here find elevated beach marks, with buried remains at all corresponding with those with which we started, there we shall find the figures to correct the theory with which we set out.

Within a few miles of the mouth of the Des Chutes River, the very evidence we need turns up. More than two hundred and fifty feet above the present level of the river, and therefore one hundred and fifty feet higher than the elevation with which we started in theory—buried in the stratified sands, clays and gravels that mark the wash of those streams into and along that old lake-beach are found the tusks, teeth and bones of the land animals of that period, marking at once the height at which these waters stood and the life record of the times. A recent visit to this locality, in company with an eminent geologist, the late Prof. LeConte, gives remarkably fresh vividness to the recollection of the facts and figures that define the position of its fossils.

A ride of four or five miles from The Dalles brings us to where three of the creeks referred to join their streams and empty together into the Columbia. The surrounding hills are composed largely of soft volcanic tufa, and through this these streams have worn deep ravines in their descent. The ravines were worn to their present depth long before the period we are describing, and when subsequently the waters rose here, backed up from the ocean, they filled these ravines, converting them into deep bays, and thus forming so many sheltered nooks into which the streams washed and in which they buried whatever the winds or floods committed to their keeping. On entering one of these ravines, we come suddenly to the edge of a newer and deeper excavation in its mid-channel. A sudden melting of snows on the neighboring hills, a few winters since, had caused these newer excavations. Scores of them were opened here within a circuit of twenty miles. The one we entered is a large one, though not the largest. It is more than a mile in length, is in some places two hundred feet wide and twenty-five to thirty deep. Along the freshly fallen sides of these new excavations one can see the distinct horizontally stratified deposits we are seeking. The record at Shoalwater Bay is the latest there, the record among these ravines is the latest here. The height of water proved to have existed so recently there must necessarily have made its mark here. And now inasmuch as these ravine sediments are the latest traces of high waters here, their elevation necessarily gives the height of those waters. And the figures that mark the height of these fossils above the present level of the river are the figures we need to complete the theory with which we started from the capes of the Columbia.

Nor is there any room for mistake here; for while this fossil sediment extends through a vertical range of more than one hundred and fifty feet, the least total altitude that will meet the conditions of the problem must take the highest portion of this fossil bed. Stating this at two hundred and fifty feet above the present level of the river is placing it at the lowest, and even then with the understanding that we are dealing with sediment and not with surface lines. Nor yet will it do to set these facts to the credit of that system of river terraces known to exist throughout the northern portion of our continent. These were described from Frazer River, by Chief Justice Begbie of British Columbia, and years ago Prof. Dana described those from Oregon and California; still later The American Journal of Science designated them as "Part of a system of terraces that covers a large part of North America north of the Ohio and existing on all streams, as far as examined, nearly to their heads in the mountains."

Now our facts and these—exclusively inland facts—refuse to be classed together. The system of old shore-lines we are tracing belongs primarily to the sea shore. These other terraces run inland, high among the mountains. The facts upon which our theory was based were gathered at Shoalwater Bay, were controlled entirely by the level of the Pacific Ocean, and scarcely affected by the flood levels in the river, and still less by any extended lake system of the interior.

Anchitherium Jaws and Teeth. *Plate XXII*

And now, with our amended theory in mind, as a measuring rod, let us retrace our steps to the lower country—the Willamette Sound of the olden time. Let the fall of the Columbia River, from this lake-shore east of the Cascade Mountains to the mouth of the Willamette River, be stated at eighty feet. Our fossil remains on this lake-shore are two hundred and fifty feet above the present level of its waters, making a total of three hundred and thirty feet as the depth of those waters above the present surface of the mouth of the Willamette River. How naturally one looks to the currents of such a vast body of water as the agency competent to the heaping up of that long sandy ridge, one hundred feet high, through which the river has cut its way at Swan Island, north of Portland. But let us follow it still farther inland. Over where Portland now stands, these waters were three hundred and twenty-five feet deep; over Salem one hundred and sixty-five feet; over Albany, one hundred and fifteen feet; over Tualatin Plains, one hundred and forty-five feet; over Lafayette, one hundred and seventy feet. A narrow strait, over the present valley of the Tualatin River, ten or twelve miles in length, opened westward upon a broad, beautiful bay, extending over the present sites of Hillsboro and Forest Grove, to Gales Peak, among the foothills of the Coast Range. The subsoil of the fine farms of that rich agricultural region, is itself the muddy sediment of that bay. Farther south over the central portion of the present valley, and lying obliquely across the widest part of the Willamette Sound, there arose above those waters an elevated island. It extended from a point south of Lafayette to one near Salem, and must have formed a fine central object in the scene. Three or four volcanic islands extended, in an irregular semicircle, where Linn County now is; and the islands of those waters are the Buttes of today—Knox's, Peterson's and Ward's. One standing on the summit of either of these buttes, with the suggestions of these pages before him, could so easily and vividly imagine those waters recalled, as to almost persuade himself he heard the murmuring of their ripples at his feet—so sea-like, the extended plain around him—so shore-like, that line of hills from Mary's Peak, on the west, to Spencer's Butte, on the south, and only lost on the east among the intricate windings of extended slopes among the foothills of the Cascades. How natural would seem to him this restoration of one of geology's yesterdays!

The shores of that fine old Willamette Sound teemed with the life of the period. It is marvelous that so few excavations in the Willamette Valley have failed to uncover some of these relics of the past. Bones, teeth and tusks, proving a wide range of animal life, are often found in ditches, mill-races, crumbling cliffs, and other exposures of the sediments of these waters, and often within a few feet of the surface. Did man, too, live there then? We need not point out the evidences of increasing interest the world feels in facts that tend to solve the doubts that cluster around this natural inquiry. A few more mill-races dug, a few more excavations of winter floods, more careful search where mountain streams washed their trophies to their burial under still waters, and this question may be set at rest, as it regards the Willamette Sound. Oregon does not answer it yet.

The three papers that follow, viz., "The Rocks of the John Day Valley," "The Development Theory" and "An Indian Legend," are here introduced because of the side light they cast on the subjects of these pages. They are papers written for lectures or publication in the early period of the writer's work.

CHAPTER XII.

THE ROCKS OF THE JOHN DAY VALLEY

In the controversies of the day on the Origin of Species, any record of the past as authoritative as that of a good geological field, covering an extensive range, and filled with minute details of events, can hardly fail to be instructive.

The basin of the Columbia River with its tributaries, offers such a history to the world, at once continuous and authoritative, reaching in its field of operations from the Rocky Mountains to the Pacific Ocean, and, in the time it covers, from the Cretaceous period to the recent. It covers even the laying of the foundations of the country, and defines the narrow strips of land that first emerged from the ocean to become the framework of the great mountain chains. Later, as the elevation and extent of the land increased, the ocean water that first occupied the depressions between was displaced, and fresh water took its place, brought there by the now greatly increased flow from the land. Henceforth, history written by the ocean ceased; history written by lakes and rivers commenced in the storing away of specimens of tree and beast, and bird, and their effectual

1 Distal of Radius, Touchet Horse. -
2 Anchitherium Molar Teeth.

Plate XXIII.

preservation as material facts in an unerring record. The sea, thus excluded, never returned to the region east of the Cascade Mountains. A vast lake system took its place, and began at once to make as well as to write its own history.

There are many residents of the Pacific Slope who will remember having journeyed from The Dalles on the Columbia to Canyon City, among the Blue Mountains. For sixty miles or more the road passes over volcanic materials, which have drifted there from the Cascade Range. Twenty miles further, and this outflow thins out into a mere capping of basalt on the hill tops. The hills themselves, and the foundations on which they stand, are here found to be sedimentary rock, wonderfully filled with the abundant records of former animal and vegetable life. Oldest of all in sight is the old ocean bed with its teeming thousands of marine shells as perfect today in their rocky bed as those of our recent seashores, their cavities often filled with calcareous spar or chalcedony, as if to compensate for the loss of their own proper marine hues. Next in ascending order come the fresh water deposits of the earlier Tertiaries, so full of the leaf prints of the grand old forests which during that age of semi-tropical climate, covered those lake shores.

The marine rocks form the outer rim, or shore line, of what was in those early times a lake, of irregular outline, extending from Currant Creek

Hill on the west to Canyon City on the east, and from the hills north of the John Day River to the Crooked River Valley on the south. Within this lake depression whose former muddy sediment is now elevated into chalky hills, so despised for their alkaline waters and unproductive soils, the geologist feels at home. How strangely out of place a score of palm trees, a hundred yew trees, or even a bank of ferns, would seem here now. And yet here these once lived, and died and were buried; and beautiful beyond description are their fossil remains even now, as they are exhumed. Seen from the summit of Currant Creek Hill, its western border, this vast amphitheatre of lesser hills presents a wild, wonderful grouping of varied outlines and colors. A spur of the Blue Mountains—its nearest point, forty miles away—covered with a dense forest, forms the dark background of the view. The varying shades of brown which characterize the older marine rocks rise in vast border masses, almost treeless and shrubless, in an inner irregular circle, while the lighter shades that fill the deeper depressions of the central portion mark the later sedimentary deposits; and then like vast ink-blots on a painting one sees here and there a protruding mass of dark colored trap. Through the heart of this wild region winds the John Day River, running westward until it passes the middle ground of the picture and then turning northward to join the Columbia. This stream, so insignificant in appearance, has done wonderful work among these hills. The river itself was, in the olden times merely a series of connecting links between a chain of lakes that extended from the Blue Mountains to the Cascades of the Columbia. It has for unnumbered ages gone on excavating vast gorges and canyons as all other streams in Central Oregon have done, till lake after lake was drained off, and their beds laid dry, stripped of enduring moisture, and slowly changed to a treeless desert. The deep excavations that resulted could hardly fail to lay bare important records of the past, cutting as they do through the whole extent of the Tertiary deposits. In a deep canyon through which runs a branch of Currant Creek, may be found the remains of the fan palm, with abundant remains of a beautiful fern, gems of their kind which no thoughtful mind can see without wonder and admiration. In another ravine are seen in great numbers the remains of a yew, or yew-like tree, that sheds annually, not its leaflets, but its branchlets; for in this form they are found of almost uniform length and structure imbedded in the rocks. This tree was evidently abundant upon these ancient shores, for it can be found at almost every spot where a little stream washed its miniature delta into the lake. Oaks, too, and occasionally a fine impression of an acorn, or acorn cup, are found at intervals from this place to the Blue Mountains.

But the great geological importance of that old lake depression does not arise from the fossil remains of its forests—beautiful, varied, and abundant as these are—but from its finely preserved fossil bones. Two species of rhinoceros lived their quiet indolent lives among the reeds that lined that old lake shore. A little beyond the southern spur of that distant mountain there evidently emptied a stream of some size, for its delta is strewn with fragments of silicified bones. Among these, bones of the rhinoceros are frequent; but the remains of an extinct animal, allied in some characteristics to the camel, in others to the tapir family, are most abundant.

Paleontologists have designated the genus by the name of Oreodon. The remains of three or four species of this animal are found in Central Oregon. One of these new to science, was discovered thirty miles from here, and was named by Dr. Leidy, Oreodon Superbus, from its superior size. The shaly rocks in which these remains are found are very brittle; and the inclosed fossils partake of that brittleness to such an extent, that, if not handled with the utmost care, they crumble into small fragments. Two nearly entire heads were discovered in a ravine that opens into Bridge Creek Valley. They had been exposed all winter to rain and frost and were very brittle—almost ready to drop to pieces. They were passed by until the following day, when a careful treatment to several coats of good flour paste was rendered the more efficient by additional pasting on of common paper. This was kept on for awhile, when it was carefully washed off and a more permanent preparation applied. These specimens now make a very passable appearance. Mute historians are they of the far distant past, uniting with hundreds of others to tell strange stories of the wonderful wealth of forest, field, and lake shore of that period. A tapir-like animal to which the name Lophiodon has been given, lived here, too. His remains indicate an animal of the size of the living tapir. Not far from these last were found some bones of a fossil peccary, of large size. Another of the denizens of these ancient lake shores bore some resemblance to the horse. The remains of this animal, the Anchitherium, were first discovered in the Tertiary rocks of France, a few years ago; more recently they were found in the Bad Lands of Nebraska, and later in the John Day Valley.

But the richest chapters in the history of the horse, in Oregon, are not from these rocks of the lower valley; for another and a later record in the upper part of the valley contains these.

Doubtless both portions of the valley were once continuous and formed one lake, but a stream of lava from the Blue Mountains seems to have run into it near the present site of Camp Watson, dividing it into an upper and lower lake. The lower one seems to have drained off first, the upper one remaining a lake into the later Tertiary period and receiving into its archives the remains of the animal types of a later age. The river was apparently turned northward by the outpouring of volcanic materials; and cutting for itself a new channel in the deep canyon, thirty miles or more away, formed a great bend and excavated an immense basin in these nearer and lighter colored Tertiary rocks. Above that bend, that canyon, and that volcanic outflow, the valley opens again; and there extending from Cottonwood Creek to Canyon City, are the remains of the upper lake depression of the John Day Valley.

This later lake depression received into its sediment a larger amount of volcanic ashes and cinders than the lower one did. Several of its strata are pure volcanic ashes, rough to the touch as pumice stone, which must have fallen on that lake in vast quantities. The purest was evidently that which had fallen directly into the lake; the less pure that which, first falling on surrounding hills, had subsequently drifted from them by the action of the winds and waters and became part of the lake sediment. Upon the hills that overlooked these lake shores, there lived three or four different species of the horse family. Their remains are easily distinguished; for the teeth are well preserved, and the teeth of the horse are well marked. Almost as well marked as these equine remains were some teeth that apparently represented a member of the camel family, found there, too, in a fine specimen of a lower jaw, silicified completely, and in solid rock. Fossil remains of other species, too, giving a wide range of life record, were found.

But the most remarkable thing about this upper lake record is that which reveals the way in which the history of this period was brought to a close. The last rock of the series fills the place of a cover to the volume. Never was cover better defined, nor more distinctly separated from the well-written and well-illustrated pages it serves to protect. The cover itself too, has a history worth reading. It extends for miles, varying but slightly in thickness, which amounts to twenty or twenty-five feet, and is throughout so entirely volcanic as to leave no room for mistake. Its materials are volcanic ashes and cinders; the cinders ranging from an inch across downward to the minuteness of the ashes. One can hardly look at a piece of this rock, without recalling the younger Pliny's vivid description of that shower of cinders from Mount Vesuvius, from which he saw people escaping with pillows tied on their heads for protection. Such showers fell here, certainly, over hundreds of square miles and in such vast bulk, that, pressed by the hydraulic force of later masses above it into a solid plate of rock, it now in this form measures from twenty to twenty-five feet through. No wonder it closed one of the finest life-records of that remote period, and with the record that volume—becoming at once the proximate cause of the changes that followed and the upper cover of the volume it sealed. But this violent destruction of the life of the period did not destroy that lake depression; it only partially filled its shallower portions and added thirty feet or more of sediment to the rest. The lake remained and still continued to receive into its archives of hidden sediment, tokens of the forces at work among the hills around it. One remarkable change marked that transition: the laboratories of the hills seem thereafter to have lost the power to send forth from their secret recesses heated vapors laden with mineral materials, capable of changing everything they touched to stone. The old sediments of that lake, if originally clay, are found changed to argillaceous rock; if sand, changed to sandstone; if washed gravel, they are found cemented into conglomerate. The new sediments, if clay, remained clay; if sand, remained sand; if gravel, remained so, unaltered even now. Long after that heaviest deluge of ashes had settled down into permanent rock, a new chapter was opened in the life record of these lake shores. The stratified materials that received these later records were washed from either shore into remarkably uniform slopes toward the middle line of the lake depression. These slopes were evidently once continuous along both sides of the valley; but since the lake was drained off by the deeper wearing of its outlet, every little stream from the surrounding hills has

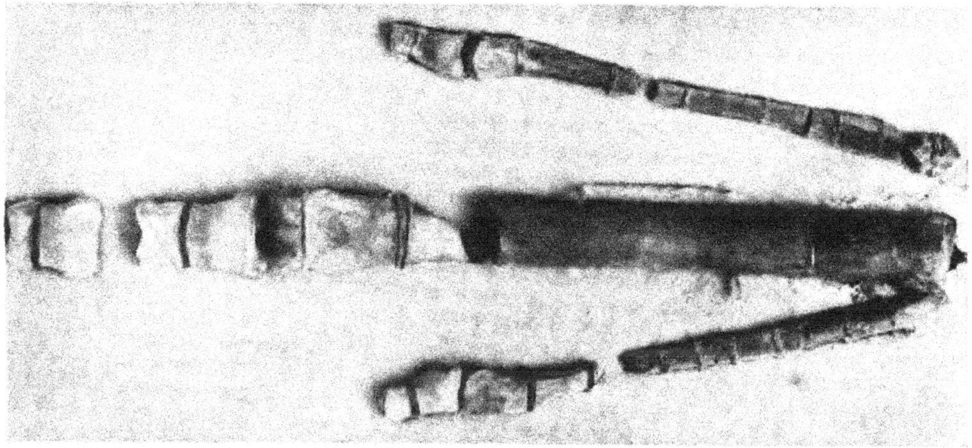

Hipparion Foot. *Plate XXIV.*

cut its own ravine through these stratified sands, gravels and clays, until what was once continuous is now cut up into remarkably uniform series of ridges, whose summit outlines stand in fine perspective, as far as the eye can reach. In the ravines that separate these ridges, the gold of this region is found, and in the diggings that result, the bones, teeth and tusks of the elephant are often uncovered, a few of which have been preserved. In the loose materials that form these ridges the closing annals of that remarkable lake-period of Central Oregon may be read as in a book. The last facts noted there are the records of the mammoth, the horse, the ox and their contemporaries.

We have thus attempted to give four or five glimpses into the grand old panoramic life record of the past in Central Oregon—successive day and night glimpses along the shores of a series of lakes that once occupied the valley, now depressed, through which meanders the John Day River.

The first one of these views is characteristic of the old marine life of the original sea bed. It is made up of a number of patches of sea beach, strewn with shells, a tooth or two, of some extinct reptile; a vertebra of another, and the marine record closes. The shoals on which these marine animals lived became elevated into the frame work of the future Oregon, while in the depressions between them her earliest historic records at once began. Oregon's Eocene, Oregon's dawn! Strange, beautiful, coincidence of fact with system! The next glimpse we get is of the Middle Tertiary period. It is distinct enough to enable us to recognize upon those lake shores the rhinoceros, the oreodon, the tapir; and then closes abruptly to give place to a record of fire and violence—the fire of the volcano, and the violence of the earthquake, bringing upon the life of the period a blotted, illegible night record in its history.

But another dawn came then; and we see, among the forms that move along those shores, the familiar ones of the horse and the camel. Again the legible record closes, and thirty feet or more of ashes and volcanic cinders cover the land, and choke and poison the waters.

A long dark and nearly illegible part of the record follows, during which no life-history was written; but during which the old throes of violence seem to have passed away, and the laboratories of the earth seem to have lost the power of forcing heated vapors to the surface, capable of changing all they touched to stone.

The mammoth, the horse, and the ox appear in the light of the dawn that follows this long geological night; and not fire, as before, but frost, seems to have closed the record marked by their fossil remains.

This alternating of light of life, and darkness of death, as read in the rocks of that region, leaves us long periods of its chronology unwritten, save by fire and flood. What are these blanks in that life-record? Have the materials upon which they were originally written been partially or wholly destroyed, or washed away? No, for in a neighboring mountain 1,500 feet in vertical section they still remain, protected by a heavy capping of basalt. The pages are there, but they are defaced by fire and ashes. But were there not, or at least might there not have been, vast periods during which no record was made? This supposition, too, is inadmissable. A lake existed here through the whole Tertiary period; and a continued lake depression surrounded by elevated ridges of hills, rising in many places into mountain magnitude, implies the deposit of continued sediment, and this necessarily becomes the page upon which the history of the life along its shores is written. The winds would always blow into the waters of the lake their burden of leaves, and the floods of winter wash there some fragments of the bones of the animals that characterized the period. It must have happened, then, that at the close of each great period, as indicated here, the animal life of these ancient lake shores was entirely destroyed by fire, flood and the poisonous vapors that tainted earth, air, and waters, or else those to whom migration was possible, escaped to some other region. The supposition of their entire destruction encounters this difficulty: the destruction of the entire fauna of Oregon, and even of the whole western slope of the continent, would not have secured the results observed, unless we suppose a like destruction extending to the Atlantic Coast, for the same animals lived there, when they lived here. Their remains are found, even to identity of species from Nebraska to New Mexico. It is difficult to assign their destruction there to the same causes that destroyed them here; or to any cause operating at once, over a whole continent, while the climate remained unchanged, and food continued abundant. On the other hand, the supposition of the escape of a portion from these destroying agencies meets, among others, this difficulty. When here, in this John Day Valley, quiet had been again restored, and the hills again clothed in verdure, and the waters had precipitated not only, but covered out of sight, their vast strata of volcanic ashes, then animal life returned too, but not the same that had previously existed. The whole fauna was changed, and even where the same type was restored as in the case of the horse, it is in some new species; the old had passed away, and forever.

If anyone supposes that all the difficulties that beset these lines of inquiry and research rest only in the path of the theologian who claims a separate creation for each great type of animal life, he greatly misapprehends the present state of these investigations. But it was no part of the plan of this article to advocate any existing theory or to

Hipparion Molar Teeth. *Plate XXV.*

start a new one in this difficult field of inquiry, so full today of conflicting views, but rather to call attention to the importance of the Columbia basin as a field filled to an extraordinary degree, with the very facts needed to throw light on the question of the origin of species. Three great ranges of mountains, and several minor ones were elevated across its water sheds, making so many immense dams, holding back the waters in extensive lake depressions, among which the river itself was for ages, but a series of connecting links.

It is now almost certain that vast lake depressions continued from their first formation, to be such, until the bones of the modern horse, ox, and elephant were received into their sedimentary deposits; thus including records covering nearly the whole period of ancient mammalian life upon the earth.

Add to these facts, that all the rocks through which the streams of this region, during this long geological period, have been wearing their way, were those of the later and softer materials, and therefore the more rapidly worn down, not only in the canyons of the larger streams, but in the ravines of the smaller ones, and upon every hillside, and we have a combination of favoring conditions, such as must make its geology accessible, very full, and important.

Indeed one can hardly look over its historic archives of the Tertiary period, without a conviction that this Columbia Basin is destined yet to be the great battle ground of conflicting theories upon the question of the origin of species.

CHAPTER XIII.

AN INDIAN LEGEND

There is an Indian legend that many years ago the Columbia was navigable for canoes, from the Chinook villages at the mouth of the river to The Dalles; that voyagers on their way up and down passed under a great natural bridge which spanned the river at the present Cascades. The great falls of the river in those days, were a little above The Dalles, at the present falls of the Tumwater, where the waters then descended perpendicularly twelve to fifteen times the height of a tall man. These falls prevented the salmon from going above that place and the Indians of the interior came there to purchase fish for which they gave buffalo and buffalo robes. But after many years, Mt. Hood and Mt. St. Helens had a quarrel, during which these mountains threw out fire and smoke, and hurled great rocks at each other. Such was the violence of the contest that the ground shook for miles around and in this commotion the natural bridge that spanned the river at the Cascades was broken down and the ruin of this bridge so obstructed the stream as to dam up its waters at the Cascades, raising the river above so high as to almost take way the falls of the Tumwater.

A two-fold result followed this violent change. The salmon, thenceforth, were able to pass The Dalles, and the Indians of the interior were no longer compelled to come to the Tumwater for fish, while the canoes of the Chinooks were thenceforth shut off from the river above the Cascades.

The gold hunter takes a pan of sand and shakes it violently in water till he sees the gold it contains, if there be any. Let us treat this Indian legend to a like process. Is it supposable that any existing cause could have increased the obstructions in the river here at the Cascades so as to have brought about the change indicated in the legend? Or are there now any existing indications of any such change having occurred?

The characteristic tendency of basalt to vertical fracture, especially as in this case, when overlying a softer rock, renders the portions of these mountains that overhang the river bank liable to a good deal of crumbling. That this friability is deep and not a mere surface movement of debris, is proved by a general but slow movement of both banks toward the river. The able engineer of the O. S. N. Co. once assured the writer that this slow glacial-like movement of the mountains toward the river was such as to necessitate frequent readjustment of the railroad lines of the Company on both sides of the river. One readily sees how this pressure of foot hills toward the bank may be unequally resisted or accelerated by the rate of erosion over

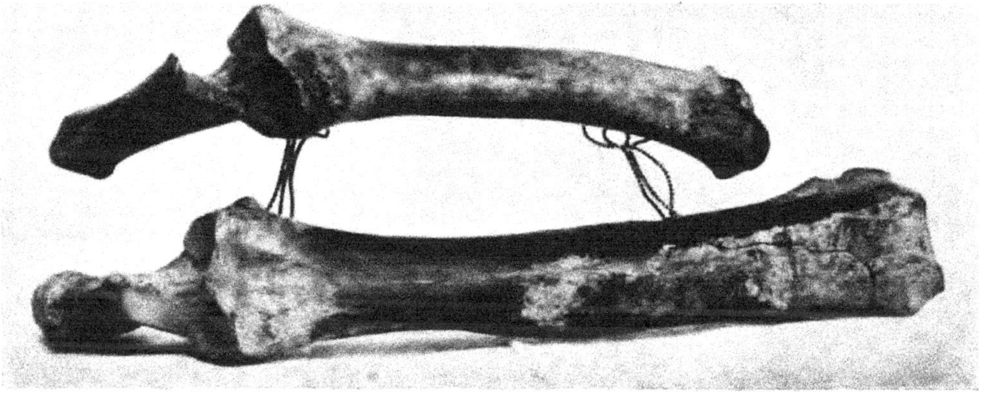

1 Ulna and Radius of Camel.
2 Same of Dray Horse.

Plate XXVI.

the river bed, or the setting of the current towards one bank rather than the other. One sees, too, how any violent disturbance, such as an earthquake would greatly increase this streamward movement, even to the extent of obstructing the channel and retarding the current so as to cause the river above the Cascades to rise above its former level.

The five miles of rapids we now call the Cascades, have a total fall of thirty-seven feet. If thirty feet of this were, by any cause, now transferred fifty miles above to the other fall at the Tumwater, the result would certainly be a barrier to all further progress upward of the salmon of the Columbia. That something has occurred here to raise the Columbia above the Cascades within a century or so, is almost certain, for submerged groves of trees occur along both sides of the river above for a distance of twelve or fifteen miles. These submerged forests are not petrified as sometimes stated, but trees in slow process of decay in positions to which landslides could not have brought them, and in a depth of water in which they could not grow. They are now in the lowest stages of water standing in a depth of fifteen to twenty feet. It is therefore almost certain that when these groves of trees were living, the Columbia River between the upper Cascades and The Dalles was more than twenty feet lower than it is today.

A strong confirmation of these facts and conclusions may be found in the United States Railway

Survey, Vol. 6, where Dr. Newberry writes as follows: "The river from The Dalles to the Cascades is very deep, has an imperceptible current and has rather the appearance of an elongated lake than of a flowing stream." Surely this looks a little like a damming up of the waters, but Dr. Newberry continues: "At intervals over the entire distance from the point where we entered the mountains to the Cascades, the river is bordered on either side by the erect but partially decayed stumps of trees, which project in considerable numbers from the surface of the water. This has been termed the 'Sunken Forest,' and has been generally attributed to slides from the sides of the mountains which have carried down into the bed of the stream the standing trees. This phenomenon, however, is dependent on a different cause. As I have mentioned, the vicinity of the falls has been the scene of recent volcanic action. A consequence of this action has been the precipitation of a portion of this wall, bordering the stream, into its bed."

This lake-like level of the waters above the Cascades, together with its great depth; this extended sunken forest whose trees are too uniformly perpendicular to permit the supposition of landslides where they stand; this glacial-like lateral pressure against the Cascade railroad, so capable of great acceleration in times of volcanic action, these facts indicate some truth in that old Indian tradition of the Lower Columbia.

CHAPTER XIV.

THE DEVELOPMENT THEORY

INTRODUCTORY

Lecture delivered in Portland, February, 1883

Reasoning *a priori*, we assume that organisms, both plant and animal, have been created by development from preexistent forms, because it agrees with the general course of nature. All the events in geology as in physics and astronomy being due to the operation of natural laws, it is reasonably supposed that the production of all the species of plants and animals from original simple forms like the monera or bacteria have been the results of natural law. The study of the early forms of life found in the Palaeozoic strata; the laws of the succession of types; the correlation existing between the development of the individual and of the members of the class to which it belongs; the parallelism between the formation and differentiation of the land masses of the globe and the successive extinctions and creations of plants and animals—all these facts, notwithstanding the imperfections of the geological record and the fact that many of the older forms of animals were nearly as much specialized as those now living, tend strongly to prove that on the whole the world as it now exists has been the result of progressive development, one form coming genetically from another; the animal and plant worlds constituting two systems of blood relations rather than sets of independent creations." Dr. Packard, Zoology, 671-2.

EVOLUTION

The doctrine of theistic evolution: that is, the doctrine that declares evolution to be God's process of creation, is now taught by all the higher colleges of our country. Among its teachers it enrolls the names of Dr. McCosh, of Princeton University, to represent the Presbyterians; Prof. Dana, of Yale, to represent the Congregationalists; Prof. Packard, of Brown University, to represent the Baptists. It claims among its seats of learning, Harvard, Dartmouth, Cornell, Michigan University, University of California and many colleges of lesser note.

Although it has this standing among religious teachers, two classes of thinkers still persist in lecturing and sermonizing the public into the belief that the doctrine itself is essentially atheistic. The two classes are, the atheists and the faithful among the theologians. That an atheist should want to usurp this doctrine is not surprising: that an orthodox theologion should help him to succeed is a matter of surprise and deep regret. If it concerned the atheist or the theologian alone the evil would be less than it is. As it is, the real harm is done to thousands of young people who read enough to know that science accepts some sort of evolution and who hear from Sabbath sermons that evolution and godliness are not inconsistent only, but destructive of each other. If these things are stated with anything like fairness to the facts, it ought to be a service to the cause of truth to spread before the public the evidence of the extent to which the development theory has already passed into general acceptance; and also to attempt to show that this acceptance, where accomplished through intelligent weighing of its truth, has in it no tendency to atheism.

Such is the aim of these pages. No effort will be made to defend or to condemn accepted views—only to record them and their results.

But a few years ago light, heat, electricity, chemical reactions and mechanical motion were supposed to be due to entirely separate acts of creation. It is now clearly seen that these and other physical forces are only separate links of one chain of underlying natural force. It is demonstrated that nothing of this underlying force is ever wasted. The motion of a mill, of an arm, of a steam engine, occurs because heat, or some other link of the chain is changed into motion. The motion thus created expends itself by becoming again heat or electricity, or some other form of the same chain of forces. Nothing of all this is now made or destroyed, not even wasted.

These things are now the commonplace facts of science. The natural effect of them on human thought would be that, whereas we once thought God created light alone, we now know he must have created a wider fact of which light is only a part. And with scientific Christians, this was the only effect the change produced. Would that it had been left to this.

How this view of the truth could lessen anyone's adoring reverence of the Infinite Source of all this wider force, and profounder power, it is difficult to understand; that it should carry with it a tendency to atheism is incredible, for somewhere in that long chain of sequences the Creator's power must come in. The normal effect upon our belief would be expressed by such a statement as this: "I once believed God created a small fact; I now see he must have created a whole system of facts at once."

This tendency to wider, more generalized facts is the one characteristic of recent scientific experiments. Our thoughts must be adjusted to this current of things if we would keep our theology a working power among men.

Still more plainly is this wider generalization marked in the domain of chemistry. In chemistry, as in other departments of science experiments continually reveal other and wider facts and forces underlying those which are already evident. The discoveries of late years through the use of the spectroscope have added greatly to this conviction. These show that the distant stars are composed of chemical elements like those of our own earth. This certainly gives one a sufficiently generalized idea of the nature of the materials out of which sun, moon and planets are made. If we consider these materials as we find them in the rocks around us we shall find evidence enough of development from simple elements to complex combinations.

As a surface fact nothing can be more simple than a piece of chalk; yet if you examine it closely you will find its simplicity to vanish and in the place of that simplicity a most complex combination of chemistry, history and mineralogy. It tells of the lowly life of a company of animals existing in the deep regions of the ocean, milleniums ago, extracting the carbonate of lime from the waters around them and through the wonderful chemical forces of life, converting this lime carbonate into bony skeletons which on the death of the animals were consigned to the deep oozy bed of the ocean to become chalk. It tells of a subsequent elevation of this ancient chalk bed into a mountain mass of a neighboring continent. How far from simple, either in time, in place or in chemistry is this strange mixture of rock and of history!

Yet you may say of this piece of chalk "God created it." So he did, but how? Evidently by a long process of development from simpler elements of time, of force, and of material to what you now find it.

A piece of granite from the hills, no more than those hills themselves, can now be regarded as a thing created in its present form by an instantaneous exercise of divine power. If you examine it closely you will find it to be a combination of three other combinations. It is made up of quartz, feldspar and mica: the quartz is a combination of silicon and oxygen; the feldspar is a combination of silicic acid and aluminum, with either lime, potash or soda; the mica is a combination of silica and alumina, and for a third element, either potash, magnesia, lime or even iron. Surely here is development in its most marked form; development through combinations complex and varied to present results. And this without at all carrying the argument to the molecules of a stage anterior to these, or to the atoms of a still more remote stage; and yet the changes this material underwent in these preceding stages, are as truly a part of the creation of your piece of granite as the combination of its quartz, feldspar and mica. How plainly then is it true that the creation of a piece of limestone or of granite consists in such a combination of atoms and forces in nature as shall secure these resulting masses, and that time, and often a good deal of it, enters into such combinations, making their existence itself a history of changes.

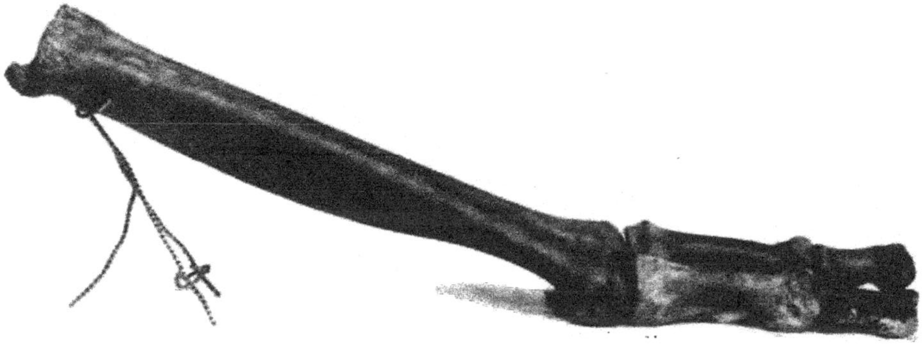

Leg and Foot of Auchenia. *Plate XXVI*

Let it be remembered that the object here is not to impress the hearer with a fact in chemistry or in mineralogy, but with the fact of a creation through development accepted throughout Christendom for the last hundred years or more, by the religious of all parties, and without any known tendency to atheism. It is not easy to see why the wider act of creation should have less need of a creator than the narrower one, or that these general systems of nature should have any less need of a plan and a designer than the more special ones of our older thought. There is in both the same need of a creator. The wider systems as well as the narrower ones will show their missing links. What matters it where these missing links occur?

> "From nature's chain whatever link you strike,
> Tenth or ten-thousandth breaks the chain alike."

If the power and wisdom of a creating God stands behind that missing link, it makes no difference where it occurs, whether in the creation of a piece of granite itself, or of the atoms of which it is composed. And yet one might say in passing, of these missing links in all systems, that there is no obvious gain in making the argument from design to hang rather on the absent than upon the present link, as if a broken link in a chain should be made to commend the skill of the mechanic more than if it were unbroken.

Several years ago the astronomer, LaPlace, published a hypothesis of development applied to the solar system, in which the claim was made that the sun, moon and planets were not created one by one from nothing by divine power; but that the matter of which they are composed once existed diffused through space; that this matter was drawn together by the mutual attraction of its particles; was condensed into such position as to give play to its chemical forces; that this condensation developed motion in the whole mass, causing it to revolve around its center of gravity; that the increase of this motion resulted in the casting off of its outer portion into space which outcast mass would in time become planets, and moons, circulating around the central mass, and that our present sun is the great central residue of all this.

With the great mass of the world's educated thinkers this explanation of God's mode of creating the solar system has passed from the domain of hypothesis to that of accepted theory; one capable of explaining facts in its connection not capable of explanation on any other theory. No one now thinks of objecting to it on the ground of its giving a substitute, in its explanation, for the power of God and thus promoting atheism, although this was often done when the hypothesis was first published. You need not be reminded that this nebular hypothesis is one of development as applied to the solar system.

But the field of scientific work which of all others shows the most marked change in the use of the word "create" is that of geology. But a little while ago it was generally believed that mountains were created as such and from nothing.

It is now clearly seen that the mountain's mass once existed in different form, perhaps extended over a plain or sea bed, occupying the very place now occupied by the mountain; and that at an earlier period the materials that now constitute its rocky mass existed as a plastic mass of mud carried to this very place by neighboring rivers.

Or if we take for our illustration a still more specific case. It is strictly correct to say God created this mountain west of Portland just where it is. But what do we mean by created here? Let us inquire. In a careful examination of the mountain itself we find at least four different kinds of materials entering into its structure. Its surface to the depth of several feet is covered with a rich bed of soil; under this surface soil is a series of beds of boulder clay; under this a varying mass of basaltic lava evidently the remians of not one, but many successive lava floods, and underlying all these a continuous mass of shales and sandstones reaching back under the mountain to Tualatin Plains in one direction, to Scappoose and the Lower Columbia in another, and to Eugene and the upper Willamette in still another. And so connected are all these parts that no portion can be separated from the others in the part it took in this one act of creating power which we call the creation of this mountain.

Each of these portions of this mountain has a history of its own—the whole a common stretch of history. If we turn to the first, the surface layer, and ask it of its record, we shall be informed of ground up material of freezings and thawings, of oxidations and deoxidations, of additions from decaying leaves and logs, by all of which means this surface soil was brought to its present condition of usefulness. If we examine the layer underlying this surface one, that of the boulder clay, we shall find a like stretch of history to mark its preparation for its place. The drift of the glacier that brought its heavy boulder masses to their present place, the history of the occasional piece of granite that seems so far out of place among porphyries and basalts as to suggest an iceberg journey from some northern shore—all these fragments of story unavoidably come in to our conception of the creation of the boulder drift that constitutes the second layer of our Portland mountain. Then we shall find ourselves in the presence of the basaltic layers of what were once, without doubt, great lava flows over level causeways long since eroded by deep valleys or narrow ravines until the direction of their outflow seems now almost incredible.

We now reach the basement sedimentary rock upon which all this upbuilding has been erected. You will find it in a few places cropping out in the river bank at low water. It extends back under the mountain to Tualatin Plains; it forms the foot-

1-3 Mylodon Toes.
2 Mastodon Tooth.

Plate XXVII

hills around the plains, it borders the valley forming its foothills through South Tualatin, Wapato Lake, Amity, Albany, Eugene and back down the valley on the east to Portland again. It is an old sea bed. Throughout its extent its fossil remains are well marked, and definitely fix the time in which its sea shells, its star-fishes and its sharks lived at home in its waters. The materials of sand and mud out of which these sandstones and shales were made, were brought here from higher lands, so that at the very foundation of our mountain base we find ourselves looking back to an earlier period for a part of the agencies that make up the history of our mountain mass. Such, in brief, is the natural history of this mountain west of Portland.

Now, if we say, as we have an undoubted right to say, "God created this mountain," I am compelled by the facts of the case to define "created," as developed through a long continued series of changes in which heat and frost, sea and land, stream and flood and tide, all did their share.

We reach a like conclusion if the object of our study be the natural history of some river channel. Look for illustration, at our own Columbia. Of this, too we may say, and say properly, God created it, the whole of it. But what does this act of creation imply? Let us see: The long winding stream of water we call the Columbia River is a vast thread that binds into geographical oneness regions wide apart and strangely varied, but united in this one tie of an extended water course.

Similar to this is its place in time: here, too, it becomes a thread that ties together widely dissimilar chapters of geological history. Let us try to recall two or three of these.

That we may get a glimpse of the first of these restorations of past history, it is requisite that we imagine the stream of time rolled back one hundred thousand years or more.

This done and we shall find the water shed of the Columbia River of that period occupying in the main, the same region it does now, and yet, along its whole course it will seem wondrously changed. It was then in its lower or western portion a broad, winding strait, bearing the same relation to the interior that Fuca Straits do now to Puget Sound. A broad beautiful bay extended southward from this strait to where Eugene now stands, fringed with deep inlets, into which mountain streams poured from the same valleys these streams now occupy. This broad stretch of inland water, let us call the Willamette Sound.

Another, and far greater extension of the Columbia River, stretched from where Walla Walla now stands to the Yakima Valley, making here, too, an extensive inland sea.

Still another extension reached from Snake River to the westward to and including the present Klamath Marsh. No facts in the natural history of the country are plainer than the evidences of these former extensions of this great water course.

Nor was this the beginning; far from it. If now we take another step into the great past we shall find still the same Columbia River, but now only as a connecting series of links between frequent lakes large and small. A river whose banks were covered with palms, whose lakes and streams were frequented by the rhinoceros, the wild horse dwarfed and giant, the tapir, the camel and many stranger forms long since passed away.

At the time of this earlier chapter the present Willamette Valley was through many broad, open straits, in communication with the sea.

If now with this extended view of its past history each epoch of which helped to form the succeeding one, we say God created the Columbia River Valley, we of necessity are held to imply through a long development of forms and materials and forces like these now at work around us. That among these were the heat of internal fires, the frosts of unnumbered winters, changes of level and changes in living forms,—which, of course, amounts to an acknowledgment that we regard the development theory as defining God's process of creating the Columbia River.

What is here said of mountain mass or of river channel, applies with like force to the creation of a whole continent. Our own continent began its history, as such, eons ago as a long strip of elevated sea bed extending westward and northward from our present Labrador. To this nucleus were added through long periods of time, and by the natural agencies of flood and tide and the life and death of plant and of animal, successive strips of land, each strip having at once a separate history of its own and a wider historical connection with the whole continent. So plainly is this true that the whole geological history is now conceded to be a marked instance of a grand system of development with plan and purpose in its movements, although conducted through ages of change and the agencies of nature.

Age after age thus left their record until our continent reached its present southern extension in Florida and the Gulf Coast, and its western extension along the shores of our Coast Range of the Pacific. To these successive areas added to the continent geologists have applied the names Laurentian, Silurian, Devonian, Carboniferous, Triassic, Jurassic, Cretaceous, Eocene, Miocene, Pliocene, Post-tertiary, each one of these names representing in succession a period of time and an area added to the continent. They stand toward each other in three lines of relationship. First—They stand in the relation of parts to a whole continent, each part in such relation to the whole that it could not possibly fit anywhere else. Second—They stand to each other and to the whole in the relation of succession in history. The third relation is that of

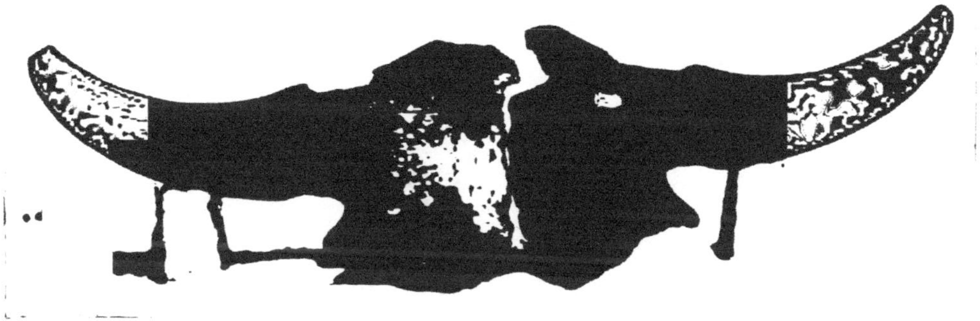

Bos Latifrons (Broad-faced Ox). *Plate XXIX.*

derivation. As before stated, each successive addition to the continent was an off-shore sea bed near the former land, the muddy sediment of which had been for ages wearing from the uplands; had been carried to the sea by neighboring rivers and distributed by tides and currents; had at length been elevated into dry land to form another, an added field, to our continental farm. Thus it was that each added area was derived from the eroded surface of its parent country, adding the relation of direct derivation to the other ties that bind the whole into one great natural development. Of course in all this, whether taken from the history of a hill, a river, valley, or a continent, we are treating of inorganic forces and developments; life as yet has not come into the question.

And now, before we carry our subject beyond the line that separates the organic from the inorganic, let us sum up our conclusions drawn from this part of our subject:

With our minds directed especially to the truths of chemistry we may state that it was once believed that God created granite directly from nothing. The educated world now believes that God created the ultimate atoms and the forces that governed their relations; and that these acted on by their surroundings, made the granite. The difference is surely this: The older belief ascribes to God the creation of innumberable separate facts; the newer thought ascribes to God the creation of a system that results in these facts.

The system among scientists prevails, for all modern researches tend towards system. Only an added evidence that God works by system.

If we direct our inquiries to the domain of geology a similar result follows. Educated men find the evidence overwhelming that God did not create at once the whole continent as men once believed, but that He so directed natural agencies and materials that the natural forces of these acted on by their surroundings developed a continent.

Here, too, the change into a generalization. God did not stop to create a single fact; He created a vast system of facts.

And as in both these departments of thought, that of chemistry and that of geology, the religious world has accepted these changed views without conscious detriment to religious faith, it must follow that development as a form of creation, as God's process of creation, in at least some departments of His work, is an accepted doctrine and is not atheistic.

But let us now go back a little in our geological history of the continent. It will be remembered that the development of our continent was described as progressive elevations of new portions of sea bed, like so many added fields to an old farm. If now we add the statement that each of these annexed fields was in its turn stocked with plants and animals suitable to its period, we have the opening of the other half of our subject, development as applied to plants and animals.

It will be remembered that there were designated ten or more of these annexed fields during the whole period of geological history. All through this history running through millions of years there were, side by side, two kingdoms of life, that of the plant, the vegetable kingdom, that of the animal, the animal kingdom. During all this time the interrelations of field and flora and of field and fauna were such that each province of each kingdom fits where it is and would fit nowhere else. This triple relationship suggests a wider system to which these parts are essential in time, in rank and in unity to the whole.

The perception of this great geological system of plants as well as of animals, long since suggested to Prof. Agassiz what was known as his system of evolution, an essentially embryonic one, and therefore not dependant on surroundings.

The other type of evolution is that of variations, promoted by surroundings and is best represented by the Darwinian system. Neither of these attempt to account for the origin of life itself. They are both content to ascribe this to God.

The scripture texts that relate to the introduction of life into the world are the following:

THE DEVELOPMENT THEORY

1. "And God said let the earth bring forth grass, the herb yielding seed, and the fruit tree yielding fruit after his kind."

2. "And God said, let the waters bring forth abundantly the moving creature that hath life."

3. "And God said, let the earth bring forth the living creature after his kind, cattle and creeping thing, and beast of the earth after his kind."

If these passages simply assure us that our Heavenly Father created life upon the earth and in the waters around the earth by starting its streams from a germ which He caused the waters or the land to bring forth, then the believer in special creations and the theistic evolutionist have here a common ground.

Beyond this, their views separate, the evolutionist claiming that God created the possibility of the whole system in its first germ of life, and so consigned it to the development of the natural world. To him, then, these passages from Genesis open a vision of a vast stream of life, beginning millions of years ago in the dawn of the Paleozoic, increasing naturally as it flowed on through the successive additions to the continent, rapidly enlarging as it flowed through the early Tertiary, till the extending continents were overspread with life in the wonderful variety of its higher forms with which the later Tertiary prepared the world for its present.

If these passages of Genesis open to us in vision this grand procession of the life of the past, our thought of God will certainly kindle no less honor to Hm, while it will be more true to the facts.

Man's place in this vast stream of life, the theistic evolutionist finds no difficulty in defining. His flesh is of the earth earthy: his animal life belongs with the broad current above described. But God assures us that He created another system, in this wider life stream, even a spiritual one; for it is written, of man, that "God breathed into his nostrils the breath of life, and man became a living soul."

Here the Christian evolutionist finds the latest and highest creative work of all—that to which all rightly tends; that to which all else was intended to be tributary—the evolution of the religious destiny of mankind.

Suppose now the question were asked, what effect will the development theory have on the faith of Christendom? It might be presumptive to attempt to give a direct answer to so grave an inquiry, but if we consult history for parallels from the past, these may help us form a judgment. About three hundred years ago a new hypothesis of the solar system was published by the astronomer, Copernicus. His theory was adopted by Gallileo and demonstrated by the help of his newly invented telescope. But the church was alarmed and asked the question: "What will become of the faith of

Christendom if these unscriptural views of the sun and the earth be generally accepted?" The poor astronomer of the telescope was condemned for heresy and compelled to retract his published convictions on pain of the penalty due to heresy. The heresy triumphed. Europe accepted the new views, but did not give up the faith of Christendom. This is certainly a case in point and ought to have its moral for us. Years passed and a new scientific heresy was published—that of the great antiquity of the earth—six thousand years would not cover the scope of history geologists saw in the rocks. The theological cry was again raised and in almost the same inquiry: "What will become of the faith of Christendom if these views are accepted?" Well, time passed, the longer chronology was generally accepted and the faith of Christendom seemed rather to improve under the change.

But yet another strain was in store for the relation between theology and science. The evidence of several lines of scientific inquiry seemed to point to a longer human antiquity than the received one. Again the old cry was raised of atheism and infidelity against the innovators, and again the newer views prevailed without much apparent change in the faith of Christendom. That these periodic conflicts between theology and science have been entirely harmless, no well informed person will claim. The church cannot put herself in a position of chronic antagonism to science without harm.

But in opening out this subject, so that we may see how much of evolution we ourselves believe; and also in enumerating the evidences that our colleges are already teaching it to our youth without taint of atheism, I have done the work proposed by these pages. That the wise and good of the nineteenth century are about to let these doctrines make atheists or even infidels of the rising generation, I cannot believe; that they are necessarily destructive of faith I believe as little. That the American church may, through their help, be able to cast aside a good deal of worthless teaching, and rise to a higher plane of working power, is far more legitimate to the signs of the times.

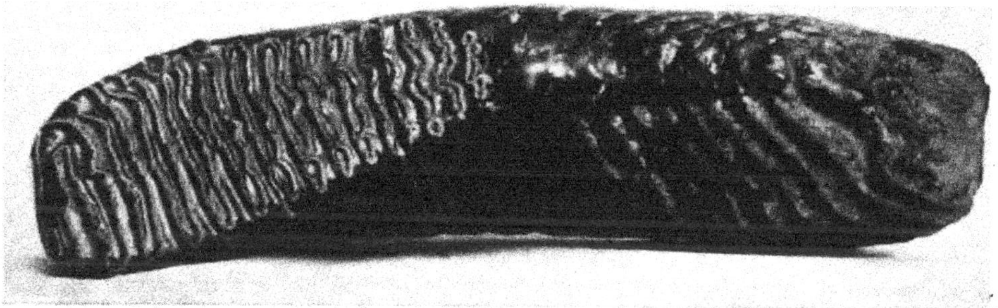

Mammoth Tooth. *Plate XXX.*

INDEX

INDEX

INDEX

INDEX

THE

FOLLOWING SELECTIONS

HAVE BEEN MADE FROM

Tributes, Incidents and Characteristic Sayings

Published in the Memorial Bulletin

Issued by the University of Oregon in 1907

"Professor Condon is widely known as a scientist, but he was more than a scientist. He was by endowment a poet. His mental powers, uncommon in other respects, owed much of their splendid efficiency to that strong yet delicate imagination which lent a charm to all he did or said. It was nearly impossible for him to be commonplace, even for a moment. His utterances fell naturally into a unique form, gentle fancy investing with poetic atmosphere even the things of every day. 'Come in,' he would say to his friend, 'when your cup is effervescing, and let us enjoy the overflow.' "

By living habitually in the higher reaches of thought his sensitive nature took on more and more attributes suggesting the sublime. He impressed one as not merely a scholar and poet, but a seer. JOSEPH SCHAFER.

THE ROCKY MOUNTAIN NAUTILUS

A lucky accident took me past Professor Condon's door at the time when the case containing the Rocky Mountain Nautilus had just arrived from the Black Hills, and the venerable Professor invited me in to "help him hurrah" as he expressed it. But as the contents of the case were unpacked, and specimen after specimen, in an almost perfect state of preservation, came into view, he forgot the presence of others, forgot everything except the beauty and wonder of the opalescent objects that glowed in his hands, or the possibilities that the unwrapped packages might contain; and as he hovered over his treasures, laying one carefully here, another lovingly there, whistling all the while softly into his beard a little comfortable tune that defies reproduction, I thought that I had never before seen such enthusiasm, such rapt absorption as this man had in his work. He straightened up once and a rare smile lighted his face as he came over to me, and laying his hand on my shoulder, said, as if in explanation, "Oh, the tune inside of me is too big for my whistle." He returned to his shells and I to my class room, realizing that the message from the Black Hills must indeed have been of rare eloquence so deeply to move the soul of this high priest of nature.

IRVING M. GLEN.

THE ROCK THAT HELD A TOOTH

In the valley of the great Columbia, a rock with tooth protruding was passed by careless men in quest of wealth until there came one with sympathetic vision who caressed the rock with loving touches and carried it away from its bed of mud, hardened into stone. With infinite care he chipped away the fragments of rock until the one tooth was joined by its fellows and there stood out in clear and perfect form the skull of the ancient horse. Even careless men now admired the beauteous sculpture of nature's older days; but the inspired geologist bore his treasure to the summit of the hills, and by a subtle chemistry of the soul he clothed it anew with flesh, and from words strange to other ears he learned of the mighty lakes and forests, of the many horses and camels and of the herds of huge mammoths whose trumpetings drowned the voices of many roaring waters.

He looked away to the hills that once held those forests, to the river that drained those lakes to the sea, and then he looked up to the starlit dome above him and cried out with reverent love, "Oh, God, Thou great Creator, lift my soul to greater light!"

EDMOND S. MEANY.

TRIBUTES

Professor Condon belonged to that rare type of teacher who takes the promising student into his heart, and gives him to drink from the wellsprings of his soul, and enchains his interest through the imagination, the perception of imminent law, and the mystic power of character over mind. Absolutely sincere, simple in life and manner, gentle and alluring in speech, Professor Condon was nevertheless one of the most courageous of men.

No visitor to the University in the days when Professor Condon was in the vigor of his beautiful old age can forget the enthusaism with which he would conduct one from case to case in that wonderful room where he kept his fossils and taught his classes. lecturing, explaining, glowing with joy over the beauty and the truth of science. It was a privilege never to be forgotten to hear him describe his collections. He would take up one specimen after another and handle them with all the tenderness of a mother caressing her child.

. . . . .

He taught with power and fruitfulness. Oregon is populated with his students who perpetuate in their lives the spirit of his deep earnestness and love of truth.

C. H. CHAPMAN.

TRIBUTES

Professor Condon was a student of history, and out of his large-hearted faith in mankind he inspired confidence in the future. In public speech his power was brilliant and rare. If the word "eloquence" be considered in its deepest significance and highest reach, it belonged to him. It was an eloquence which, like that of Webster, made the loftiest subject minister to the needs of today. More than all, he was a lover of his fellow men, and he was ever a participant in the joys and necessities of those about him. No one within his sphere was external to his interest. He was a realist in recognizing the everyday problems of life that confront us all, and in counselling that they be solved with practical means that are available to common sense and industry. But, ah,—and here was a rare combination—he was an idealist of the most delicate mould. His intellectual faculties, poised, calm, seemed ever in the light of his sympathies and warm emotions; his emotions, burning with a steady flame, seemed always controlled by his judgment. His rare imagination saw daily about him a new heaven and a new earth; his eyes were so kindly, and he looked right into your nature and always saw the best in you, and somehow his face and words were a blessing.

He seemed almost to belong to another age, with that stately courtesy that fitted him like a garment. He stood within the doorway of the home and gave the kindly, deferential, courtly greeting to all who visited him. The tender affection of his friends was sweet to him and the things eternal seemed just beside him.

Those who loved him can hear the slow, reverential tones of adoration as Professor Condon would pronounce the prayer of Moses, the man of God: "Lord, Thou hast been our dwelling place in all generations. Before the mountains were brought forth, or ever Thou hadst formed the earth and the world, even from everlasting to everlasting, Thou art God."

LUELLA CLAY CARSON.

CHARACTERISTIC SAYINGS OF PROFESSOR CONDON

"The Holy Spirit is a scientific necessity, a constant emission from the Being of God, affecting human character just as the sun affects the crude starch of an unripe peach, transforming it into sugar, and making the rich, luscious, perfected fruit. The human brain has been gradually evolved to prepare it to receive these rays of divine light, and the human spiritual life is but the crowning of preparation."

"Sin is being behind in God's plan of progress—being like the tiger and the hog, when God wants all human beings to leave the animal nature behind."

"God wants, commands you to use your own judgment in the light of this twentieth century, to tell you what is right and beautiful and true. I believe in inspiration as a living force now."

One of several volumes of different books of the New Testament was found in his room at the University, and in explanation he said: "I like to have two or three of them scattered in my room, for nothing at all but for their fragrance."

"The vision of things to be done may come a long time before the way of doing them appears clear, but woe to him who distrusts the vision."

GLOSSARY

Argillaceous, containing much clay.

Calcareous rocks, the limestones.

Canine tooth, dog tooth, sharp pointed corner teeth.

Carboniferous, the great coal period.

Carnivora, flesh-eating animals.

Carpus, the bones of the wrist.

Cartilaginous, having gristle or cartilage in place of true bone.

Champlain, post glacial.

Comprehensive type, the origin of several lines of life.

Conglomerate, a rock composed of pebbles cemented together.

Cretaceous, the chalk period, close of Reptilian age.

Deployed, diverged in development.

Digit, a finger.

Environment, surroundings.

Eocene, the dawn of the recent.

Equus, Latin for horse.

Escarpment, a precipitous side of hill or rock.

Femur, thigh bone.

GLOSSARY

Friability, easily reduced to powder.

Functional, useful.

Glacial age, the age of ice.

Incisors, the cutting teeth in front part of the jaw.

Infiltrate, to enter by permeating the pores.

Inflorescence, covered with alkali salts like flowers.

Interstratified, stratified between other formations.

Jurassic, the middle period of the Reptilian age.

Mammals, animals which suckle the young.

Marsupials, like the opossum and kangaroo.

Metacarpus, the bones in the palm of the hand.

Metamorphosed, rock altered by heat.

Miocene, the middle Tertiary.

Molars, grinding teeth.

Moraines, lines of rock and earth fallen upon glaciers.

Oxidation, the union of oxygen with other elements.

Paleobotanist, a student of fossil botany.

Paleontologist, a student of ancient life.

Paleozoic, ancient life.

Phalanges, bones of fingers and toes.

Pleistocene, just preceding recent time.

GLOSSARY

Pliocene, more recent than the Miocene.

Preglacial, before the age of ice.

Radius, the exterior bone of the fore arm.

Recent, from the close of Pleistocene to the present.

Ruminant, an animal that chews the cud.

Sedimentary, rocks formed by water sediment.

Silicified, to be permeated by silica or quartz.

Tertiary, the three periods, Eocene, Miocene and Pliocene.

Tibia, the shin bone.

Triassic, the oldest part of the Reptilian age.

Ulna, the larger of the two bones of the fore arm.

Spotted something that isn't quite right?

Let us know and we'll fix it for future reprints.
And as a thank you for your diligence, we'll
send you a free Sothis book of your choice.

publisher@sothis.press

This page intentionally left blank

Made in the USA
Coppell, TX
05 November 2026

11307333R00163